# *Ladybugs and Lettuce Leaves*

*Prepared by Project Inside/Outside
Somerville Public Schools*

*A Publication of Center for Science in the Public Interest
Washington, D.C.*

# LADYBUGS AND LETTUCE LEAVES

**Readings and Activities in Gardening, Environmental Education and Science**

a publication of
**CENTER FOR SCIENCE IN THE PUBLIC INTEREST**
1755 S Street, N.W.
Washington, D.C. 20009

Written, illustrated and produced by
**PROJECT OUTSIDE/INSIDE**
Somerville Public Schools
Somerville, MA 02143

Funded under an ESEA Title IVc grant from the Massachusetts Department of Education, the Comprehensive Employment and Training Act, and the Somerville Public Schools.

Special thanks to Judith Dortz, Raymond Izzo, Joseph Pignatiello, and Ellen Sarkisian for their ideas and support throughout this project.

No part of this publication may be reproduced, photocopied or transmitted in any form without written permission from the Center for Science in the Public Interest.

© 1982 Center for Science in the Public Interest

**Cover illustration:** Patrice Moerman

ISBN 0893290963

| | |
|---|---|
| Production/Design | John Madama |
| Writers | Jorie Hunken |
| | John Madama |
| | Pamela Pacelli |
| Editors | Dorinda Hale |
| | Sandra Hackman |
| | Michael Jacobson |
| | Carol Kouhia |
| Designers | Melinda Gordon |
| | Nick Thorkelson |
| Graphic Artists | Lyrl Ahern |
| | Jimmy Garcia |
| | David Hines |
| | Maxine Novek |
| | Adele Travisano |
| Photo Printing | Clark Quin |
| Curriculum Content | Lyrl Ahern |
| | Tracy Barnes |
| | Charles Blanchard |
| | Patricia Brennan |
| | Gordon Campbell |
| | Melinda Gordon |
| | Janis James |
| | Anne Marie Kapitan |
| | Mary Kelleher |
| | Mary Lewis |
| | Philomena Lombardi |
| | John Neals |
| | Constance Phillips |
| | Ellen Sarkisian |
| | Richard Waldman |
| | Joanna Wicklein |
| | Carol Wintle |
| Evaluation | Jean Schensal |

## Drawings

Lyrl Ahern: pp. 39, 40, 63, 68, 79-83.
Melinda Gordon: pp. 5 (sun), 27, 42, 43, 72, 73.
Jorie Hunken: pp. 69, 71, 74-76.
Maxine Novek: pp. 13, 14, 22, 35-37, 40, 46-49, 52, 53, 63, 70.
Nick Thorkelson: pp. 16 (map), 24, 26, 32, 41, 56, 59, 62.
Adele Travisano: pp. 15, 16 & 17 (vegetables), 20, 28-30, 45, 50, 51, 54, 55, 58, 61, 64, 65.

## Photos

NASA: p. 4; Mt. Palomar: p. 5 (galaxy); Harvard University, Tozzer Library: pp. 5 & 6 (hunter, Egyptian); John Madama: pp. 5 (ferns), 7, 23, 25 (ocean), 31, 38, 55, 57, 66, 67, 77-79; Julie Stone: p. 8; Hale Reservation/Lenny Myers: p. 9; Massachusetts Audubon Society: p. 10 (moth, leafbug); Robert Silberglied: p. 10 (thornbugs, stickbug); Burpee Seed Company: p. 12; Howard Harrison: p. 25 (mountain); Appalachian Mountain Club/David Hoyt: p. 25 (river); Denver & Rio Grande Western Railroad: p. 34; National Park Service: p. 60.

## Answers

Page 23: there are eleven seeds—sunflower, kidney bean, black-eyed pea, corn, cucumber, lentil, watermelon, pumpkin, beet, radish, rice.

Page 77: close-up of (a) bean seed; (b) inside a peanut; (c) banana cross-section; (d) leaf stomata; (e) tomato cross-section, showing seeds being born; (f) cross-section of root pipelines.

# Contents

1. A Journey Through the Leaves of Time ... 5
2. How Does Your Garden Grow ... 7
3. Exploring Your Space • Sharp Eyes ... 9
4. Seed Catalogues ... 11
5. Nutrition • You Are What You Eat ... 13
6. Plant Maturity • How Long Before a Plant Grows Up? ... 16
7. Climate & Plants ... 18
8. The Sun • Center of Life ... 20
9. Seed Elections ... 22
10. Hidden Treasure • Soil ... 24
11. Soil From Rock ... 25
12. Soil From Life ... 28
13. Soil From Life • Compost ... 29
14. Soil Minerals ... 32
15. Changing Environments ... 34
16. Earthworms • Giants of the Soil ... 36
17. Seeds ... 38
18. Germination • Getting Started ... 40
19. Sprouting Seeds to Eat ... 41
20. Lotus Seeds • How Long Will a Seed Last? ... 42
21. The Young Plant • Making and Finding Food ... 43
22. Old Plants in New Places • Changing Environments ... 44
23. From Inside to Outside • The Coldframe ... 45
24. From Shell to Plow ... 46
25. Companion Planting ... 48
26. Drawing Your Garden Plan ... 50
27. Sample Garden Plans ... 51
28. Planting the Garden ... 52
29. Planting Seedlings ... 53
30. Scarecrows ... 54
31. Community Gardens ... 55
32. Old MacDonald Had a Farm ... 56
33. Roots • Waterworks of Plants ... 58
34. Stems • Transporters & Supporters ... 60
35. The Water Cycle ... 62
36. The Tree Experiment ... 64
37. Leaves • Food Factories ... 66
38. A Tour of the Factory ... 68
39. Adaptations • How Plants Survive ... 69
40. Weeds • Good Plants in Wrong Places ... 70
41. Garden Insects • Pests & Friends ... 72
42. Insect Control ... 74
43. Life on a Carrot Flower ... 76
44. What is It? ... 77
45. Back to Seed ... 78
46. The Making of Fruits & Seeds ... 80
47. From Outside to Inside • Harvest ... 84

# A Journey Through the Leaves of Time

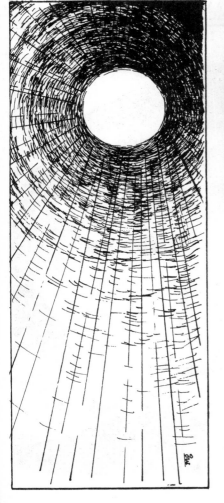

Spiral Galaxy

Ferns

Cave Painting

The earth, a giant ball of rock, water, and air, floats in space among millions of burning stars. One of those countless stars is very special. It is called the sun—a brilliant ball of fire which provides the earth with warmth and energy.

Before there was any life on the earth, the sun's rays touched and warmed the barren globe each day. Millions of years of sunrise and sunset passed before the first plants struggled to survive. The sun provided these plants with the energy to grow and reproduce. Without the sun they could not live. With the sun they covered the earth like a blanket of green.

Then animals of every kind evolved — flying, swimming, and crawling. All of these creatures, large and small, needed plants to survive.

Millions of years passed again before the first humans walked among the creatures of the earth. The first humans picked and ate the plants they found. They hunted the wild animals. They traveled across continents as they covered the earth in search of new supplies of food. The humans were dependent on the sun, the plants, and the animals for their lives.

Then a time came when people discovered how to grow their own food; how a seed, if planted in the earth, would produce a new plant. And so gardening began.

The discovery of gardening soon spread to every corner of the earth. Instead of roaming the land for food, people settled and grew the food they needed to survive.

Skilled gardening gave rise to great civilizations. From ancient Babylon to mystical Egypt, from wondrous China to the wild Americas, gardening and farming became a way of life. Slowly, the mysteries of seed, soil, and weather were learned from working with the land.

Today, you can grow your own food and learn as people did in ancient times how we depend on the sun, soil, and plants. Pass on what you discover to your friends and neighbors and become part of the history of gardening.

Egyptian Pictograph

American Indian Painting

# 2
# How Does Your Garden Grow?

People have been growing gardens, singing about gardens, talking about gardens, and depending upon gardens for thousands of years. Have you ever grown a garden? Do you think it will be easy or hard? Do you think it will take a lot of time or a little? What will you have to do to get ready? What will you plant? All of these questions are important, whether you're going to be a gardener or just find out how plants and gardens grow.

# Gardening is Both an Art and a Science

- Gardening is an **art** because as a gardener you have the opportunity to make a piece of land a beautiful — and delicious — **environment**, filled with life and color.

- It is a **science** because you have to experiment to see what works in your garden and what doesn't. You have to think about the weather, the soil, water, insects, and even the birds — all the parts of nature that go together to make up the environment.

Always remember that plants are living things and that they are sensitive to their environment, just as you are. If you care about their needs and help to meet those needs, they will cooperate with you and produce a beautiful garden. Guaranteed!

## DOING
1. With paper, pencil, and crayons, design a garden of your own. What will you grow in it? Use your imagination!
2. Share your garden plans with one another. Hang them around the room and make it bloom.
3. Write a story about a garden you have grown or one that you have seen growing. Do you have parents or grandparents who garden? Share your stories with your class.
4. Can you think of any stories, songs, or poems in which gardening is mentioned?

# Exploring Your Space
# 3
## • Sharp Eyes

The first thing a good gardener does is examine the **garden site** (site: place or location of some activity). By looking at this site you will find out what kind of life is around you now. Try to imagine how this site will change once you plant your garden.

## DOING
Can you name your five senses? Keep them alert, for if you do you will make many great discoveries.

You will need:
- a pencil or pen
- some paper
- a thermometer

1. Go to the site and write down the things that you see. Get down on the ground and look around you. Feel the soil. Look up at the sky and notice the trees. They may be bare now. Will they spread their leaves in the summer and cover the garden with shade?
Listen closely for the sound of the wind, a bird, an insect. Look for different colors.
2. Take the temperature of the ground using the thermometer. Take the temperature of the air. Record the date and the results in your journal.
3. Take a photograph of the garden site.
4. In the classroom, after you return from the site, make a class list of the things everyone found. Copy the list and the date in your journal.
5. Every month, go out and observe the site again. Don't forget to take the temperatures and a photograph each time. As the months pass, you will notice how your environment changes. By the end of the school year you will have discovered many wonderful things.

# SHARP EYES

Nature makes many creatures blend into their environment. In this way they are protected. Look at the pictures on this card. What can you discover in them? Use your **sharp eyes** to find out!

Thornbugs

Moth

Leafbug

Stickbug

# Seed Catalogues

4

Many people look through books called seed catalogues before they begin to garden. These books show pictures of hundreds of different types of fruit or vegetables that you can grow. Seed catalogues can help you decide what seeds to buy. Looking through seed catalogues can also help you discover some plants that you never knew existed.

## DOING

Choose a partner and go through a couple of seed catalogues together. Pick out some of the vegetable and flower seeds that look good to you. Try some you've eaten before and, as an experiment, choose some new ones. Pick six kinds of vegetables and three kinds of flowers you want to plant.

| Vegetables | Flowers |
|---|---|
| 1. | 1. |
| 2. | 2. |
| 3. | 3. |
| 4. | |
| 5. | |
| 6. | |

You have just finished the first step in making a garden: you have chosen what you would like to grow. There are other things, however, to think about before you decide with your class which seeds to buy and plant. These are:

**Nutrition**

**Sun/Shade**

**Climate**

**Time to Maturity**

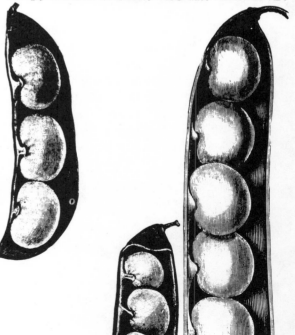

# Nutrition • You Are What You Eat

## 5

In olden times, farmers grew plants not only to eat, but also for their ability to cure certain diseases. Plants were given names such as Heart's Ease, Heal-All, Boneset, and Eyebright. The plants were named this way because the farmers felt that these plants helped their bodies to remain healthy. The health-giving properties of plants were discovered over hundreds of years of experimentation.

What early farmers didn't know about (at least in the way we do today) was the science of nutrition. **Nutrition** is the study of food and how it affects our health. To remain healthy, our bodies need a daily supply of **vitamins**, **minerals**, **proteins**, **fats**, and **carbohydrates**. These are called **nutrients**. The vegetables and even the flowers in your garden contain these nutrients. Without them you would become sick, and if you don't eat enough of them, perhaps you are not as healthy as you might be.

On page 14 there is a chart of some of the most important vitamins and minerals. The list tells which vegetables are the best sources of these nutrients. It also explains what each vitamin and mineral does for your body.

### DOING
1. Work with a partner and design a garden which contains all of the vitamins and minerals in the chart. Present it to the class and compare gardens.
2. Look at your list of vegetables on p. 12. What vitamins and minerals do they contain? Are you concentrating more on some vitamins and minerals than on others? What vegetables could you add that would even out your list?

## Vegetable and Fruit Nutrition Chart
— Serious vitamin deficiencies are rare in the United States and Canada —

| VITAMIN | BEST SOURCES | WHAT IT DOES FOR YOU |
|---|---|---|
| A | Pumpkins, Winter squash, Sweet potatoes, Collard greens, Carrots, Kale, Broccoli, Spinach, Red peppers, Watermelon, Tomatoes | Keeps eyes healthy and capable of seeing well at night (prevents night blindness) |
| B | Broccoli, Peanuts, Peas, Turnip greens, Lima beans, Brussels sprouts, Collard greens, Dried beans and peas | Keeps skin, eyes, and hair healthy. Helps nerves function properly. Helps the body make blood. Enables you to use the energy found in carbohydrates (bread, potatoes, pasta). |
| C | Green peppers, Collard greens, Brussels sprouts, Oranges, Strawberries, Tomatoes, Grapefruit, Canteloupe | Helps heal wounds. Helps make cementing materials that hold body cells together. Keeps gums healthy. Fights infection. |
| D | Not found in vegetables, but if you work in the garden you'll get plenty of it from **sunshine**. It's called the sunshine vitamin; eggs and milk also contain this vitamin. | Helps calcium make strong bones and teeth. |
| E | Vegetable oils, Green leafy vegetables | Helps keep membranes surrounding cells strong and healthy. |

**MINERALS**

| | | |
|---|---|---|
| Calcium | Collard greens, Broccoli, Kale, Swiss chard, Dried beans | Works with Vitamin D to build bones and teeth. |
| Iron | Broccoli, Peas, Beans, Leafy green vegetables | Helps blood carry oxygen. |

## Food for the Health of It

In the last section you learned that vegetables give your body many of the nutrients it needs to stay healthy. In this section you will learn that there's more to good nutrition than just eating fresh vegetables. Vegetables are some of the best sources of vitamins and minerals but vegetables alone can't keep you in top shape. To look and feel your best you also need to think about other foods that you may or may not be eating.

Today, many foods that Americans eat every day can be bad for your health if you eat too much of them. These foods are usually high in sugar, fat, salt, or chemical additives. They are sometimes called "junk" foods. These foods can cause you to be overweight, have poor skin, get tooth decay, or develop heart disease when you get older.

If you want to be at your best, find ways to avoid these types of food and replace them with fresh whole foods. Use the chart on this page as your guide to healthy eating.

## DOING

1. In your journal, write down in a column all the foods you ate yesterday for breakfast, lunch, dinner, and snacks.
2. Then, look at the Food Guide and mark a large "X" next to the foods on your list that are under Foods To Avoid.
3. Next, make another list of all the foods marked with an X. Next to each of these foods write a food from the "Eat More" column that you might eat instead.
4. Try to make these good food changes whenever you can. You'll not only look and feel better but will discover the great taste of fresh foods.

## SNACK FOOD GUIDE

| Eat less of these ... | ...Because they are: | Eat more of these |
|---|---|---|
| Soda pop, fruit drink, milkshakes | High in added sugar | Unsweetened fruit juices, unsalted club soda, fruit juice mixed with club soda, water, lowfat milk, milk-banana-fruit shake |
| Candy bars, doughnuts, cakes, pies, cookies, pastries | High in added sugar and fat | Fresh fruits (bananas, apples, berries, melons), blueberry or bran muffins, whole grain breads, bagels, or rolls (spread with apple butter of fruit flavored yogurt) |
| Bacon, hot dogs, cold cuts, processed cheese | High in fat, salt, and additives | Baked or broiled fish, chicken, or lean meats, yogurt, cottage cheese |
| Potato chips, corn chips, cheese puffs, french fries | High in fat, salt | Unsalted popcorn, pretzels, nuts & seeds; roasted soybeans |

# Plant Maturity • How Long Before a Plant Grows Up? 6

What do you think **maturity** means? Does it refer to the size of a plant or person? Is maturity measured by the way something looks or acts? Look it up in a dictionary and then discuss it with the rest of the class.

Being in school makes it important for you to think about when your vegetables and flowers will **mature.** Do you want to **harvest** everything before you leave for vacation in the summer, or do you want to arrange to tend the garden a little longer so that you can have some of the late maturing vegetables?

**DOING**

1. Choose 10 vegetables and list their planting dates for your zone.* Then figure out the dates each will mature. List which vegetables will be ready for harvesting in May, June, July, August or September.

*Use the map to find out your area's zone number. Then, if you live in Zone 1, use the earlier planting dates from the chart below. If you live in Zone 2 use the later planting dates. Those living further north than Zone 1 or in higher mountain areas should ask a local garden store. Those living in Zone 3 should plant 4 to 6 weeks before the earliest date given.

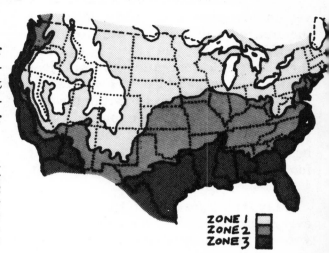

ZONE 1
ZONE 2
ZONE 3

**PLANT MATURITY GUIDE FOR UNITED STATES**

| Vegetable | Planting dates for zones 1 & 2 | Days to Maturity | Climate | Maturity Date |
|---|---|---|---|---|
| Beans | 4-25 to 6-1 | 60-80 | Warm | |
| Beets | 4-1 to 6-1 | 55-80 | Cool | |
| Broccoli | 4-15 to 6-1 | 60-75 days from seeding transplant | Cool | |
| Brussels Sprouts | 3-25 to 5-1 | 90 days from transplanting | Cool | |
| Cabbage | 3-15 to 4-10 | 65-100 days from transplanting | Cool | |
| Carrots | 4-10 to 5-1 | 65-75 days | Cool but don't mind warm | |
| Cauliflower | 3-15 to 5-1 | 55-65 days from transplanting | Cool | |

| Vegetable | Planting dates for zones 1 & 2 | Days to Maturity | Climate | Maturity Date |
|---|---|---|---|---|
| Chard | 4-15 to 6-15 | 60 days | Cool | |
| Collards | 4-1 to 6-1 | 80 | Cool-warm | |
| Corn | 4-25 to 6-1 | 65-90 | Warm | |
| Cucumbers | 5-15 to 6-15 | 55-65 | Warm | |
| Eggplants | 5-15 to 6-15 | 80-85 days from transplanting | Hot | |
| Kale | 4-1 to 5-15 | 60-80 | Cool | |
| Lettuce | 4-1 to 5-15 | 40 days | Cool | |
| Melons (Watermelon, Canteloupe) | 5-15 to 6-15 | 75-120 days (depends on type) | Hot | |
| Onions | 4-1 to 5-1 | 95 | Cool | |
| Peas | 3-10 to 5-10 | 55-70 days | Cool | |
| Peppers | 5-15 to 6-1 | 60-75 from seedlings | Hot | |
| Potatoes | 3-15 to 5-15 | 70-90 | Cool | |
| Pumpkins | 5-15 to 6-15 | 120 | Warm | |
| Radishes | 3-10 to 5-15 | 21 | Cool | |
| Spinach | 3-1 to 5-1 | 45-50 | Cool | |
| Summer Squash | 5-1 to 6-10 | 60-65 | Hot | |
| Tomatoes | 5-5 to 6-15 | 65-85 days from transplanting | Hot | |
| Turnips and Rutabagas | 3-10 to 5-15 | 70 | Cool | |

# Climate & Plants

Polar

Temperate

Did you know that 10,000 years ago a mile-thick sheet of ice covered all of New England? And long before that palm trees grew on the land that is now near the Arctic Circle. Why isn't this true today?

Today scientists are trying to find out why the weather patterns change slowly over thousands of years. Some believe that the earth's orbit around the sun changes. Others say that all the continents are slowly floating around the globe and that once North America was next to Africa at the equator.

**Climate** is the word used by scientists to describe the weather patterns that take place over thousands of years. Climate is the average type of weather found in a particular area. For instance, the desert has a dry climate, but on a certain day rain might be falling. We would say that the weather was wet on that day, but the climate of the desert is still dry because, over a long period of time, very little rain falls in the desert.

Climate is very important because it determines what kinds of plants can grow in a particular area. Some plants like a hot climate, others like a cool climate, still others like a wet or dry climate. Can you name some of the different climates pictured on this page? What types of plants grow there?

Tropical

Desert

# DOING
## Apples and Oranges:
## A Case of Two Climates

Here are two state maps, one of Florida and one of Massachusetts. Beside the maps you will find some information about the climate of each state.

Think about an apple and an orange. One will grow in Florida but not in Massachusetts; the other will grow in Massachusetts but not in Florida. Read the descriptions and then try to match the fruit to its state. Explain why you made your choice.

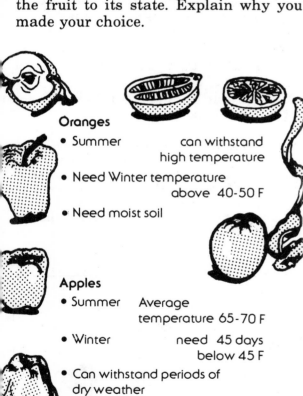

**Oranges**
- Summer     can withstand high temperature
- Need Winter temperature above 40-50 F
- Need moist soil

**Apples**
- Summer   Average temperature 65-70 F
- Winter    need 45 days below 45 F
- Can withstand periods of dry weather

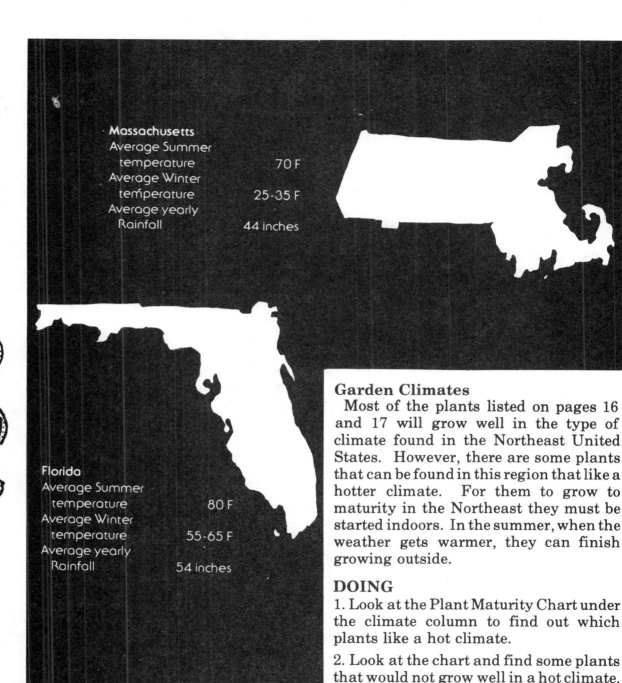

**Massachusetts**
Average Summer temperature    70 F
Average Winter temperature    25-35 F
Average yearly Rainfall    44 inches

**Florida**
Average Summer temperature    80 F
Average Winter temperature    55-65 F
Average yearly Rainfall    54 inches

## Garden Climates

Most of the plants listed on pages 16 and 17 will grow well in the type of climate found in the Northeast United States. However, there are some plants that can be found in this region that like a hotter climate. For them to grow to maturity in the Northeast they must be started indoors. In the summer, when the weather gets warmer, they can finish growing outside.

## DOING

1. Look at the Plant Maturity Chart under the climate column to find out which plants like a hot climate.

2. Look at the chart and find some plants that would not grow well in a hot climate. Record them in your journal.

# The Sun • Center of Life

We have already talked about the importance of nutrition, time to maturity, and climate in gardening. What about the sun? The sun is the center of all life. If it burned out, plants could not grow and we would all freeze to death in a very short time.

We can learn how to use the precious sunlight to our advantage. When we plant a garden, we need to remember that plants are very sensitive to sunlight. Some, like tomatoes and sunflowers, need a great deal of sun. Others, like parsley and lettuce, need shade.

## DOING

It is important to know how many hours of sunlight your garden will receive every day so that you can figure out which plants will grow best there.

1. Look at the drawings on the bottom of this page. They show how to find the number of hours of direct sunlight for any garden.

A. Stretch your hand out in front of you at arm's length. Your palm should be facing you as shown in the drawing.

B. With one eye closed, look at the amount of sky that your hand covers. This hand span, measured up and down against the sky, is equal to one hour of time.*

C. Using a compass, find east and west.

D. Face east (where the sun rises) and count the number of handspans from where you can first see the sky to directly over your head. In the drawing, the student begins counting handspans from the roof of the school building.

E. Face west and count the number of handspans of sky in that direction.

F. Add the east and west totals together. The number of handspans you counted will equal the number of hours of direct sunlight. The garden in the drawing will receive about six hours of direct sunlight.

2. How many hours of sun does your garden receive? Are there any shady spots?

3. On the next page you will find a chart of sun- and shade-loving plants. Look at it carefully and then answer these questions in your journal.

A. Look at the vegetables and flowers you chose on page 12. Which need a lot of sun? Which need less sun or shade?

B. Are there any seeds on your list which you cannot plant because the garden doesn't get the amount of sun they need?

C. List your findings and discuss them with your class.

*Note: since all hands aren't the same size you may have to adjust your handspan. To test if your handspan is accurate, measure the number of handspans from the horizon to directly over your head. If this number is six then your handspan is accurate. If not, adjust your handspan so that six will measure from the horizon to directly over your head.

## Sun & Shade Loving Plants

| Vegetables | Full Sun | Partial Sun | Some Shade | Shady |
|---|---|---|---|---|
| Beans | | ● | ● | |
| Beets & Carrots | | | ● | ● |
| Broccoli & Cabbage | | ● | ● | ● |
| Brussel sprouts | | ● | ● | |
| Cauliflower | | ● | ● | |
| Chard | ● | ● | | |
| Collard Greens | | ● | ● | |
| Corn | ● | | | |
| Cucumbers | ● | ● | | |
| Eggplant | ● | | | |
| Kale & Lettuce | | ● | ● | |
| Melons | ● | | | |
| Onions | | ● | ● | |
| Peas | ● | ● | ● | |
| Peppers | ● | | | |
| Potatoes | | ● | ● | |
| Radishes | ● | ● | ● | ● |
| Spinach | | ● | ● | |
| Squash & Pumpkins | ● | | | |
| Tomatoes | ● | | | |
| Turnips & Rutabagas | | ● | ● | |

| Flowers | Full Sun | Partial Sun | Some Shade |
|---|---|---|---|
| Aster | ● | | |
| Chrysanthemum | ● | | |
| Coral bells | ● | ● | ● |
| Cosmos | ● | | |
| Daylily | ● | ● | ● |
| Delphinium | ● | | |
| Hollyhocks | ● | | |
| Impatiens | | ● | ● |
| Marigolds | ● | | |
| Nasturtium | ● | | |
| Pansy | | ● | ● |
| Petunia | ● | ● | ● |
| Phlox | | ● | |
| Snapdragon | ● | | |
| Sunflower | ● | | |

Full sun: Plant exposed to direct sunlight at least 8 hours a day.

Partial sun: Plant exposed to direct sunlight about 6 hours a day.

Some shade: Plant exposed to direct sunlight only 4 hours a day.

Shady: Plant is in shade most of the day.

# Seed Elections ★ 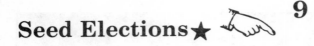 9

Join with the other members of your class and vote for twelve types of vegetables and flowers that you will be planting in your garden.

REMEMBER:   Nutrition   Sun
     Climate   Time to Maturity

**FOR YOUR JOURNAL**
In your journal, write down the date and record the list of vegetables and flowers chosen by your class. Then order the seeds your class has chosen. One package of seeds for each type of plant will be enough.

How many kinds of seeds can you find in this picture?

Answers on copyright page, (p. 2).

# Hidden Treasure • Soil

10

Look at the picture on this page. If we could cut a section out of the earth this is what we would find inside.

As you can see the earth is made of layers of solid and liquid rocks. At the center of the earth is a solid rock core. Next is a layer of liquid rock and then a layer of rock slush. Finally, and directly below us is a 40-mile-thick plate of solid rock that is moving! This plate is only one of many plates that move around the globe. The plates move because they are moving on the layer of rock slush.

Both land and water ride upon these plates. The plates can only move a few inches each year—but over millions of years they can move thousands of miles. When plates collide they form mountain ranges. Earthquakes also result from plate movements.

Lying on the very top of the plates is a thin layer called soil. Although soil is usually only a few feet thick, all life depends on it.

Soil is made of different materials blended and mixed together. One part comes from rocks and another part comes from decayed plants and animals. Let's look closer and find out how soil is made and why it is so important for life.

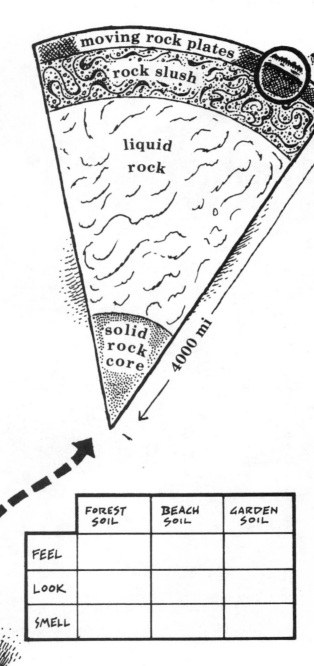

### DOING

Collect different types of soil from areas near where you live. You might also be able to have soil sent from relatives in other states.

1. Place and label each type of soil in a jar. Pick up each type. Let it sift through your fingers. Smell it. Look at it carefully. What do you see in it? Use a magnifying glass.
2. In your journal, make a chart like the one on this page and record your observations.
3. How are the soils the same? How are they different? Which ones do you think would be the best for growing plants? Why?

|  | FOREST SOIL | BEACH SOIL | GARDEN SOIL |
|---|---|---|---|
| FEEL |  |  |  |
| LOOK |  |  |  |
| SMELL |  |  |  |

# Soil from Rock

## 11

For many millions of years, weather has been breaking and carving the rocks of the earth. The rain does most of the work as it forms streams and rivers which grind and wear away the rocky surface.

As water rushes from the mountains to the oceans, it loosens small rocks and carries them along . . . pounding them against bigger ones, chipping off pieces, grinding them against one another . . . rubbing and scraping and wearing the rocks down into **sand** and **silt.**

Wind, carrying tiny pieces of sand, blasts every exposed surface. Waves crash upon the shore and wear it away.

During the great **Ice Ages,** mile deep glaciers of ice and snow crushed and powdered rocks into **clay.** Slowly, over millions of years, the rocky surface of the earth was broken down into the small pieces that are part of your soil.

Can you name and describe another way in which rocks get broken into small pieces? (Clue: Why do roads get potholes in the winter?)

Glacier Park, Montana

Lovell River, New Hampshire

Atlantic Ocean, Massachusetts

**DOING**

In this experiment we are going to **separate the three different types of rock material** contained in your soil samples. These materials, called **sand**, **silt**, and **clay**, contain the minerals which are needed by all plants and animals.

You will need:
- 1 jar with a lid
- soil samples - garden, forest, beach
- 1 large index card
- water

1. Fill your jar about ½ full of water, then add one type of soil sample until the jar is almost full. Different groups in your class can test the different samples from the garden site, forest, and beach.
2. Cover the jar and close it tightly.
3. Holding the jar with one hand on top and the other on the bottom, shake well until the soil and the water are all mixed up together.
4. Now put the jar on a table and let the ingredients settle for about an hour.
5. Take the index card and put it next to the jar. You will see distinct layers. Mark on your card where each layer begins and ends. Draw a line across the card so that it will be clear.
6. Each layer will contain a different rock material - either sand, silt, or clay. See if you can tell which is which from these clues:

**sand**: largest and heaviest pieces; grain-like with a lot of air space between each grain.

**silt**: small pieces; not much air space.

**clay**: very small, light pieces packed closely together. These pieces may also be suspended in the water layer. If you let your sample settle for a few days, this layer will become easier to see.

Which soil layer is the thickest? What material does each soil sample have the most of? Do you see any other materials in the jar besides sand, silt, and clay? What are they? Record your observations in your journal.

## DOING

1. Use a shovel (spade) to dig a cube of soil out of the earth. It should be about 6 inches on each side. Be careful and gentle; try to keep your soil cube in one piece.

2. Look for the different layers that are shown on this page. You are looking at what is known as a **soil profile**. These layers of soil were made from the slow breakdown of **rocks** and the decay of **dead plants and animals**. Can you tell which layer has more rock in it? Which has more decayed plants and animals? Why?

The soil layers and the materials in them support life by:

- providing plants with the nutrients they need to live and grow. Animals obtain these nutrients when they eat plants or other animals that eat plants.

- absorbing and holding rainwater like a sponge. Plants can absorb this supply of water as they need it.

- providing plants with a place to grow.

If you do not see all the layers that are on this card in your soil profile, it is because they were deeper than you were able to dig. You can usually see these lower layers when someone is digging a deep hole for a house or an apartment building.

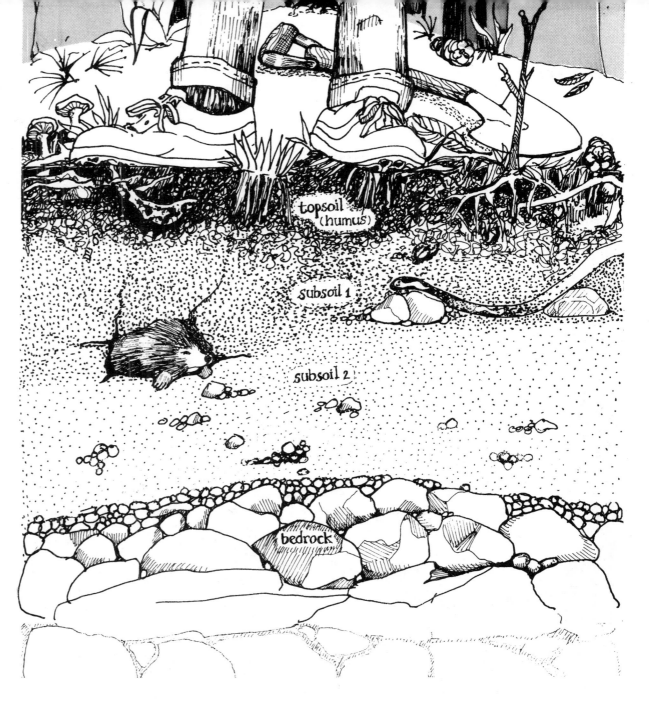

27

# Soil from Life

## 12

## Who Does All This and How? The Decomposers

fungi and molds

Have you ever noticed what happens to leaves when they fall off the trees? Or to the bodies of birds and animals when they die in the forest? Have you ever taken a walk in the woods and come across an old tree stretched across the path, with moss and mushrooms growing on it and hundreds of spiders and bugs making their home in it? If you have, then you have seen the beginings of the way in which another part of the soil is made. This part, which comes from dead plants and animals, is called the **organic part**.

The **decomposers** are countless billions of small animals and plants that live in the soil, air, and water. Many of them are so small that they can only be seen with a powerful microscope. One type, called **bacteria**, are so tiny that one spoonful of soil can contain more of these creatures than there are people on earth.

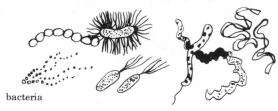
bacteria

**Fungi** and **Molds** are another type of decomposer. They are much larger than bacteria and sometimes can be seen with the naked eye. Have you ever seen an old piece of bread or fruit with mold growing on it?

The decomposers use dead plants and animals as food. When something dies, the bacteria fungi and molds which happen to be present start eating it. They eat and grow and multiply so rapidly that in a very short time millions of them are working on the dead plant or animal. It is their eating which causes what we call **decay** or **decomposition.**

When things decay, they form a material called **humus.** Humus is rich in minerals and other nutrients. The humus becomes part of the soil.

So you see that even after an animal or plant dies, it is still useful. Through the action of the decomposers, the dead animal or plant supplies nutrients for future living things.

# Soil from Life • Compost

**13**

## Making Fertilizer With The Decomposers

Each inch of soil contains billions of bacteria and fungi that are continuously breaking down all dead plants and animals into a rich material called humus. When people use the decomposers to break down food or garden waste, a material called **compost** is made. Compost is one type of fertilizer.

Compost is a valuable material for gardens; it makes the soil rich in minerals and nutrients needed for healthy plants. Compost also makes soil light and airy, which allows water and air to easily reach the roots of your vegetables. Your class may want to have a large compost pile for your garden.

## DOING
Soil From Life

You will need:
- petri dish with a cover or small cans or pie tins covered with plastic wrap.
- soil
- pieces of fruit, vegetables, bread, each no larger than ½ cubic inch
- water

1. Place a thin layer of soil on the bottom of the dish.
2. Put 3 pieces of food on top of the soil. Moisten them and the soil with water.
3. Cover the dish. Remove the cover for a few seconds every couple of days to let some air in.

## FOR YOUR JOURNAL

1. Record the day you begin and what pieces of food you use.
2. Record the day you notice a change in each of the food pieces.
3. Describe the changes. What color is each mold or fungi?
4. How long does it take before you cannot recognize each piece of food (how long does it take to **decompose**)?
5. Survey your class: how many different kinds of molds or fungi grew? What colors were they?
6. Did the same color fungi or mold grow on the same kind of food? Why?

## DOING
Outdoor Composting

You will need:
- 9 feet of chicken wire, 2 or 3 feet high
- kitchen waste, weeds, and leaves
- some loose soil
- a shovel or pitchfork

1. To begin an outdoor compost pile, first choose an out-of-the-way shady spot. Then set up a simple circular cage with about 9 feet of chicken wire. The cage should be 2 to 3 feet high.
2. Next make a compost "cake". Throw a layer of kitchen waste, weeds, and leaves into the cage. Add a thin layer of soil on top of the waste. Continue to alternate layers of kitchen waste and weeds with layers of soil.
3. Keep the pile from drying out with an occasional watering, but don't add so much water that the pile gets too soggy. Bacteria and fungi in the soil will start the decomposing process.
4. After a couple of weeks, mix the pile by turning it over with a shovel or pitchfork. This lets in air, which helps the decomposers and also allows you a chance to check on any changes in the material.
5. Every 2 weeks, mix the compost pile again. Are there earthworms present?
6. After 6 to 8 weeks, when your compost is crumbly and moist, it will be ready to go into your garden.

**The Final Test**
A soil rich in compost will crumble and have plenty of air space between the pieces. Does your garden soil meet this test?

# Soil Minerals 14

Minerals are substances that are part of all living and non-living things. Iron is a mineral and so is copper.

Good soil contains at least 14 different minerals. The growth of healthy plants requires all 14 of these elements. Minerals become available to plants when rocks erode and plants decompose.

Plants absorb these minerals through their roots. Once inside the plant the minerals become part of the plant's structure or take part in energy reactions that keep the plant growing. These minerals eventually become part of your body after you eat a plant.

Poor garden soil is often too low in one or several of these 14 minerals. Nitrogen, phosphorous, and potassium are the minerals most likely to be present in low amounts in poor soil. Plants growing in this type of soil will have stunted growth or odd-colored leaves. Poor garden soil usually results from over-using land without putting back any nutrients.

To find out if your soil is good, have a soil test done to see if it is low in any of the minerals mentioned. Your own soil test kit, a local gardening store or your state agricultural agent can help you.

If your soil is found to be low in any one of these 3 minerals you can add them back into the soil by using a material called **fertilizer.**

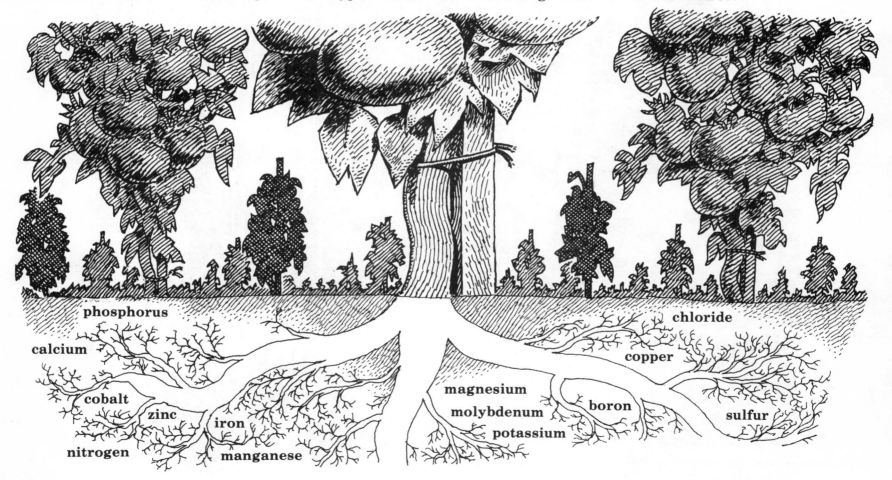

There are two main types of fertilizers: chemical and organic:

## Chemical Fertilizers

Chemical fertilizers are made in factories. They are usually expensive and can be harmful if used in excess. Their only benefit is that they make minerals quickly available to plants. If you do use them be sure to read the directions carefully.

## Organic Fertilizers

Organic fertilizers come from decomposing plant and animal material or natural rocks. This type of fertilizer is highly recommended. Organic fertilizers such as cow manure or compost not only supply a balance of minerals but also improve the soil's ability to hold water and circulate air. They are also inexpensive.

Even if your soil test reveals an adequate supply of minerals, adding a little compost or manure will be of benefit.

## pH Test

Another type of soil test measures the acid or alkaline content of the soil. This is called a pH test. PH is measured on a scale from 0 to 14. A pH reading below 7 is acid, above 7 is alkaline.

Generally soils east of the Mississippi are acid. Those west of the Mississippi are alkaline. If your soil has a pH reading below 5.9 add a material called limestone (see chart). This will raise the soil's pH towards 7 which is ideal for growing vegetables.

### Types of Organic Fertilizers

| NAME | WHAT IT DOES | MIX INTO SOIL |
| --- | --- | --- |
| cow or horse manure | minerals and soil conditioner | 25 lbs./100 sq. ft. |
| compost | minerals and soil conditioner | 25 lbs./100 sq. ft. |
| green manures (winter rye) | minerals and soil conditioner | plant in fall, turn into ground in spring. |
| peat moss | mainly soil conditioner | spread 1-3 inches on top of soil and mix in. |
| bone meal | minerals | 1 lb./10 sq. ft. |
| limestone | pH adjuster | 1 lb./30 sq. ft. |
| rock phosphates | minerals | 1/2 lb./100 sq. ft. |

# Changing Environments

15

Sometimes it is very easy to see how natural things like plants or water change the environment. For instance, when you look at a picture of the **Royal Gorge**, you can tell that water **erosion** (wearing away) has made a very deep cut into the earth. Big changes like the Royal Gorge are the result of small erosion changes occuring constantly over long periods of time. What do you think happened to all the rock that was eroded by the water?

Right now many erosion changes are taking place in your schoolyard and neighborhood. Because many of these changes are small, they are not very obvious. You need sharp eyes to see them.

Royal Gorge, Colorado

## DOING

Look in your own environment for examples of the following changes. In each case, something is being eroded into smaller parts. These small parts eventually become soil.

### Water Erosion

**Ice**: When water gets in a crack and freezes, it expands, making the crack even bigger. Eventually the once big rock, building, or stone is cracked into very small pieces that become part of the soil. Potholes in the street are good examples of ice erosion.

**Stains**: As rain water runs down the sides of a brick or stone building, it dissolves minerals that are in the stone. Often these minerals have a special color. Can you find any stains on your school building? On other buildings?

**Gullies**: When water runs down a hill, it washes away (erodes) the soil and leaves a gully. Notice how small gullies resemble the Royal Gorge. The Royal Gorge began as a small gully.

**Gravel Piles**: A gravel pile washed onto the sidewalk is a good clue that water erosion has taken place. What happens to the size of the pieces of gravel as they continue to be eroded over the years?

**Rain Gutters**: Look on the ground near rain gutters for changes in the earth's surface.

**Stone**: Look around the schoolyard for very smooth, round stones. Stones like this once were in swift rivers where the sand in the rushing water gradually wore them smooth.

### Air Erosion

**Rust**: When iron is exposed to air and water, the oxygen in the air turns the iron into rust. Iron fences or steel cans left untended outside will eventually turn entirely to rust. Soil that is very red may have a lot of iron in it.

### Plant Erosion

**Cracks**: When plants grow in cracks in rocks, buildings or sidewalks, their roots break and crack the rock into small pieces.

Draw a picture of what you think your schoolyard will look like in a million years.

# Earthworms
## • Giants of the Soil

**16**

When have you seen earthworms? While digging outside? On the pavement after a rain? While fishing with "night crawlers"? Earthworms are not very big creatures, but they do a very big job keeping the soil fit for plants. Here are some facts about earthworms which will help you understand them.

Earthworms **recycle** and concentrate decomposing organic matter and materials. They do this by eating soil! As soil passes through the earthworm's digestive system it is changed into a material (recycled soil) that plants like. Earthworms deposit this recycled soil on the earth's surface in the form of soft pellets called castings. If you see a pile of castings you can be sure that earthworms are nearby. Castings are a great fertilizer for the soil.

A tiny pile of castings doesn't look like much, but if you could weigh all the castings that earthworms deposit in one acre during one year, they would weigh 15 tons! Tiny earthworms do a giant's job fertilizing the soil.

When earthworms eat soil, they make tunnels in it. They shove their heads into the soil and push to make an opening, taking mouthfuls of soil as they go. These tunnels are also very important to plants because they allow air and water to reach the roots more easily. The earthworms, however, don't like the tunnels when they are full of water because then they can't breathe. Like us, earthworms need air. If their tunnels fill with water, they will suffocate. This is why you see them on the surface of the earth after a rain. They have come up for air.

Earthworms prefer the night because the heat of the sun will dry out their bodies and kill them. They need to keep their bodies moist. When they come to the surface at night, they eat fresh and decaying leaves that are on the ground.

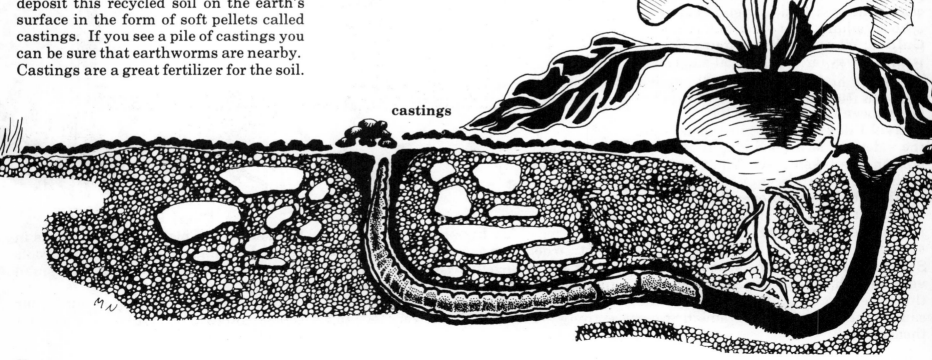

castings

**DOING**

You will need:
- vinegar
- small stick or Q-tip
- glass
- sandpaper
- paper of various colors
- container
- soil

Find an earthworm and bring it to class. Be sure to handle it gently and keep it on a moist towel.

1. Try to find these parts of an earthworm:

**Head** — The pointed end, it has a hornrimmed mouth. The eyes are small and can see only the movements of shadows.

**Clitellum** — This is a light-colored band. Two earthworms connect at this point when mating.

**Skin** — Tough and shiny, the skin is lighter on the worm's underside, where tiny hairs help the earthworm to move. Can you feel them with your fingers?

**Tail End** — The tail end is slightly flattened and muscular. It stays in the tunnel while the worm eats. If startled, the earthworm quickly pulls its body back to safety underground.

2. A worm feels with its skin. Touch a small stick to different parts of its body. Note how it reacts each time. How does a worm show discomfort?

3. A worm tastes with its skin. Dip a small stick in vinegar and touch various parts of the worm's body with it. Are some areas more sensitive than others? Can you think why?

4. What kind of surface does a worm prefer to crawl on? Try these and then think of your own: glass, a desk top, paper, sandpaper. (Return your earthworm to a moist place in between so it won't dry out.)

5. How does your earthworm react to light? Does it react differently to lightcolored paper than to dark-colored paper?

6. To find out what an earthworm eats, you need to make a home for it. Use a container that will hold several inches of soil and that can be covered to retain moisture. Put a variety of different leaves on the surface and keep a record of which ones your worm likes best. Does it like fresh or wilted leaves? How about if they are moldy? Put some of the food you eat, such as bread, candy, or fruit, on the soil. Which of these foods does the earthworm like? How much does the earthworm eat in one night?

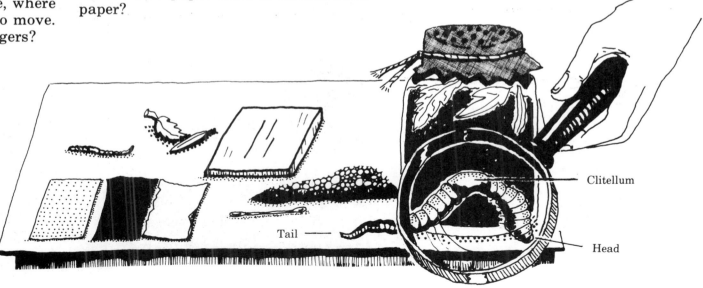

# Seeds

17

Have you ever seen maple "key" seeds flying in the wind? Or blown on dandelion seeds to make a wish? Have you ever eaten pumpkin seeds or heard of Johnny Appleseed?

Did you know that peanuts are seeds? Rice is also a seed and so is popcorn. Bread is made from wheat seeds.

In fact, every flower, fruit, vegetable, and tree comes from a seed of some kind —with each seed being different in size, shape, and color. A type of palm tree grows from a 60 pound seed while one million daisy seeds weigh only one ounce.

Some seeds float, some fly, and some burst from their parent plant like a bullet. Other seeds have hooks on them so they can catch a ride on an animal.

Although all seeds are different, they all have one purpose: to grow into a new plant.

Something to think about: "A seed is a link in the chain of life."

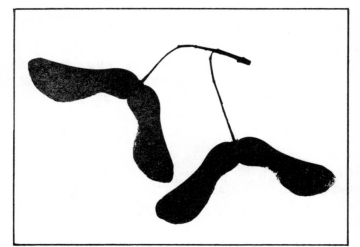
Maple "Key" Seeds

Dandelion Seeds

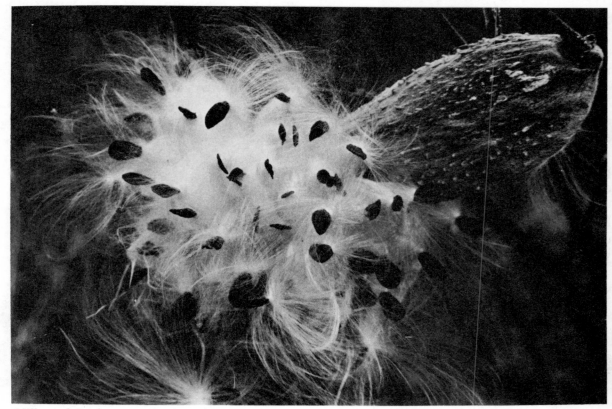
Milkweed Seeds

## DOING
1. Be a collector. Look for seeds in your kitchen, around your neighborhood, or at a park. Watch how they travel. Bring them to class and make a large seed display. Compare them for shape, size, weight, and color. What kinds of plants will your seeds become?
Plant them and see for yourself!

2. Looking inside a seed.
   A. Soak several medium to large seeds overnight in water (dried beans and peas are especially good).
   B. The next day look for special places on the seeds where you might be able to spread them open with your thumbs. Do so.
   C. Find the **seed coat**, the outer protective covering, on the two halves of a seed.
   D. Under the seed coat is a large area known as the 'seed leaves' or **cotyledons**. The cotyledons will give the young plant its first supply of food.
   E. Next, find the tiny young plant. It is very small compared to the cotyledons.

This is the part of the seed that will increase in size to become the mature plant. Look closely and you will find a pair of leaves and a part that will become the root and stem. Use your magnifying lens. Every plant, from a bean to a 300-foot tree starts from such a small begining.

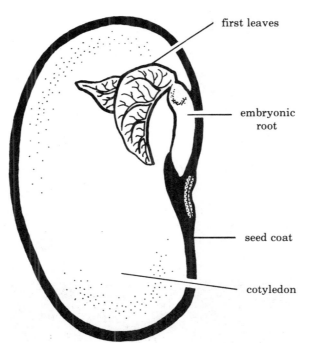

Bean Seed

## DOING
Enough Tomatoes
to Cover the Earth
   With a partner, take a tomato and split it open. Count the number of seeds in one half of the tomato.
1. How many seeds does it take to make one tomato plant?
2. How many plants could you grow from one half of the tomato?
From the whole tomato?
3. If one tomato plant produces 30 tomatoes, how many plants could you grow from all the seeds in those 30 tomatoes?
4. How many tomatoes could you get from the plants produced in question 3? (Why don't tomatoes cover the earth?)

# Germination • Changing from a Seed to a Young Plant

**How Does a Seed Sprout?**

Now that you've brought in many kinds of seeds and discovered their basic structure, answer this question: why aren't all those seeds growing? Why aren't they coming out of seed packages, busting through the Uncle Ben's Rice box, and popping out of apples?

One reason is that almost all seeds need the same basic conditions to sprout. Do the following experiment and you will learn what those conditions are. Then you will understand why the seeds aren't sprouting out of those seed packages.

Sprouting, or changing from a seed to a young plant, is called **germination**.

## DOING

**What Does a Seed Need?**

Before you begin this experiment, think about yourself for a moment. When you were born, what did you need to grow? What couldn't you live without? Think of the most basic things and write them down.

After you have finished the following experiment, discuss and compare basic human needs with the basic needs of a seed.

You will need:
- 4 petri dishes with covers
- blotter paper
- 8 seeds (bean, pea, or radish) 2 for each dish
- a refrigerator or freezer
- water

1. Label
Dish A - Water & Light
Dish B - Water & No Light
Dish C - Light & No Water
Dish D - Water & Cold

2. Put paper and 2 seeds in the bottom of each dish.
3. Dish A: Moisten the paper with water and place the dish in a sunny window. Do not allow the paper to dry out.
4. Dish B: Moisten the paper with water, place it in a dark closet or in a bag. Do not allow the paper to dry out.
5. Dish C: Do not add any water. Place the dish in a sunny window.
6. Dish D: Moisten the paper with water, place the dish in a refrigerator.
7. Observe changes in the seeds for 10 days. Make a chart like the one here and record your observations in your journal.

Discuss with other members of the class what every seed needs to germinate.

8. Look at a sprouting seed closely. Open it. Find the cotyledons, leaves, and the part that will become the root and stem.

Has the seed changed? How?

| Jar | Condition | What Happened | Why |
|-----|-----------|---------------|-----|
| A   |           |               |     |
| B   |           |               |     |
| C   |           |               |     |
| D   |           |               |     |

# Sprouting Seeds to Eat 19

Seeds of certain plants can be sprouted easily and eaten as nutritious, low-calorie vegetables. Alfalfa seeds, mung beans, chick peas, lentils, and wheat grains are among the most commonly sprouted seeds. Sprouts are crisp and delicious in sandwiches, soups, salads, and just plain. To sprout seeds, follow these directions. (Be careful not to sprout too many seeds at once. Because seeds increase greatly in size after sprouting, use about ¼ cup of seeds for every quart jar you would like to fill.)

1. Rinse seeds in a jar and soak them in warm water overnight.

2. Drain off the water and place seeds in a large jar. Secure several layers of cheesecloth or fine plastic mesh over the mouth of the jar with a rubber band.
3. Keep the jar in a warm (68° to 74°), dark place.
4. Rinse the seeds 2 to 3 times each day with warm water, leaving the cheesecloth or wire mesh in place.
5. After rinsing, leave the container inverted so the excess water will run off.

6. As the leaves of the sprouts form, expose them to indirect sunlight to increase their food value. The sprouts are ready to eat when they are 1 to 2 inches long.

# Lotus Seeds • How Long Will a Seed Last?

Once there was a beautiful lake in China. In the waters of this lake grew the sacred and respected lotus plant. Each year, seeds from the lotus fell into the clear water and slowly sank to the bottom of the lake. Over many years, the lake dried up. The seeds that had fallen from the flowers of the lotus stayed buried in the muddy bottom of the dried-up lake. The seeds were very hard, covered with a tough outer skin.

Many years passed. The land that had once been a lake was used for farming. A scientist came and began to dig in the farmland, for he was interested in its history. He found the lotus seeds. As an experiment he decided to try to sprout them, so he took them to his laboratory in Washington, D.C. He put them in strong acid to dissolve the hard seed coats and then planted them. He saved a few seeds to be tested with a new method called radiocarbon dating, which would tell him exactly how old the seeds were.

While the seeds were in the soil, he found out that they were over one thousand years old! After learning that news, he doubted that they would ever germinate. Then one morning in June 1952, a tiny sprout poked through the soil. The thousand year old seeds had sprouted!! Today they are still growing in the Kenilworth Aquatic Gardens of Washington D.C.

Of course, not all seeds last that long. The hard coats and the fact that the seeds were buried so deeply helped these lotus seeds to survive.

# The Young Plant • Making and Finding Food  21

In our experiments, we have seen how the seed provides food and a safe environment for the young plant. We have looked at the conditions a seed needs to germinate and we have studied the soil. Still some questions remain. What happens after the seedling uses up the food that was in the cotyledons? How does it survive in its new environment? How does it get the water and food it needs for growth? How will the young plant become the food that, someday, you may be eating? All of these questions will be explored in our next experiments.

## DOING
Growing Seedlings

If you remember, plants like tomatoes, peppers, and eggplants need an early indoor planting so that they can grow to maturity in our climate. We will plant these seeds now and watch how they grow. When it gets warmer outside, we'll put them in the garden.

You will need:
- small planting boxes (flats)
- soil mixture
- gravel
- pepper, tomato, and eggplant seeds
- a clear plastic bag with a tie
- a ruler

1. Place gravel along the bottom of the flats.
2. Fill the flats ¾ full with soil mixture.
3. Read each seed package to find out how deep to plant the seeds. If you plant them too deep or too shallow, they may not grow.
4. With a ruler, measure the depth at which the seed is to be planted. Mark it on a stick. Push the stick into the soil as far as the mark and drag it along in even rows about 2 inches apart. (Note: If you have very tiny seeds, they will have to be sprinkled along the top of the soil and then patted down.)
5. Drop your seeds evenly and carefully in the grooves you made with the stick.
6. Push the soil back over the rows and tamp (pat) it down gently with your fingers.
7. Water the flats.
8. Place each flat inside a clear plastic bag. Seal the bag with a tie. The plastic bag will hold in moisture so you will not have to water the seedlings often.
9. When seedlings are 1 - 2 inches high, remove the bag to better observe their growth. Remember to water more often now.

## FOR YOUR JOURNAL
1. Count the number of seeds you have planted in each box. Label each box and the day planted.

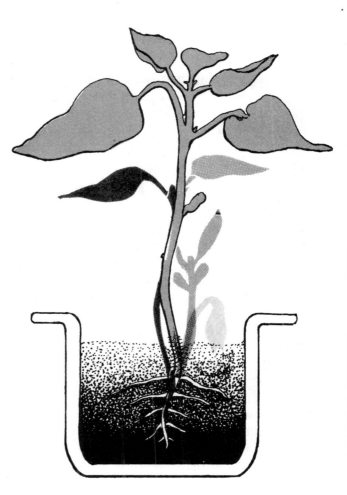

2. Record the day the seeds first germinate (the first time you see them poking through the ground).
3. When you feel that all of the seeds have germinated, count the young seedlings and compare this number with the number of seeds you planted. What was the percentage of germination?

# Old Plants in New Places • Changing Environments    22

Imagine eating a meal that contained mostly corn, squash, and beans every day. Before 1770, New Englanders had no choice. They could only eat those plants that would grow in their cool climate. Vegetables and fruits such as cantaloupe, watermelon, green peppers, and tomatoes were unheard of in New England.

In the early 1700's, people began traveling on trade missions to tropical countries. Sailors brought back tales of fabulous fruits and vegetables (oranges, lemons, pineapples, and mangoes were some of them). Sometimes they would even bring back a sample. The kings, queens, and the wealthy merchants who sponsored these trade missions were very excited about trying to grow these exotic fruits and vegetables in their homeland.

But there was a problem. The climate of the homeland was often not warm enough or wet enough to allow these plants to grow. Here's what they did.

They made a special environment for the plants where the temperature and moisture could be controlled to create a tropical climate. One example of this was the giant "Orangerie" built by Louis XIV of France, the "Sun King". It was one of the first glass **greenhouses** and was used to grow delicious and healthful citrus fruits for the palace dwellers.

Of course, most people couldn't afford to build greenhouses. They continued to eat what they had eaten for centuries: in England, peas and beans; in Sweden, turnips; in what is now the United States, squash, corn, and beans.

Today many people have greenhouses. You can see them advertised in most gardening magazines. Nurseries, plant stores, and many large food companies depend on these artificial environments to grow food during the cold months.

Even if you don't have a greenhouse you can still maintain a warm temperature in your home and classroom. For this reason, you can grow a wider variety of fruits and vegetables than early New Englanders could.

# From Inside to Outside • The Coldframe

## 23

After your seedlings have grown for a while in their pots they are ready to be set outdoors. However, since they have been pampered with warm inside temperatures they may be shocked by suddenly placing them outdoors. That is why gardeners who start plants indoors put them through a process called **hardening off.** Here is how it works:

DOING — optional
1. Two weeks or so before you plan to set your plants outside (check with your teacher or plant book to tell you when this is), put the plants in small coldframes. A **coldframe** is a small greenhouse heated by the sun. It creates a new environment in between the warmth of your classroom and the cold of the outdoors. It will allow your plants to adjust slowly to outside temperatures.

Leave your plants in their pots. On warm days open up the coldframes so that the plants do not receive too much heat. Don't forget to close them again at night.

2. Leave a few plants outside the coldframe. Compare them with the ones inside. Observe the stem, leaves, and general condition of the plant.

3. Record the temperature of the air outside the coldframe and inside the coldframe. Why are they different?

# From Shell to Plow  24

For thousands of years people have used tools to prepare their gardens. In ancient times, people used only simple objects found in their natural environment to make their tools.

**DOING**
1. Think about the things you have to do to prepare a garden. Here are some suggestions:
   You have to make rows for the seeds.
   You have to make the garden smooth.
   You have to dig out stones.
   You might have to remove grass.
   You might have to take out a tree stump.
   You have to water the plants.

Now invent at least three tools for doing these jobs, using only natural materials found outside. Share your inventions with the class.

2. Did you ever think of your hand as a tool? The human hand is the most wonderful tool there is. Just imagine how many different jobs it can do. And it never wears out!

   A. For the next day, think of your hand as a tool. Observe what you do with your hands. Then make a list of the different jobs your hands do. On page 20 your hand was used as a tool to measure time.

   B. If the human hand is such a good tool, why did people invent new tools?

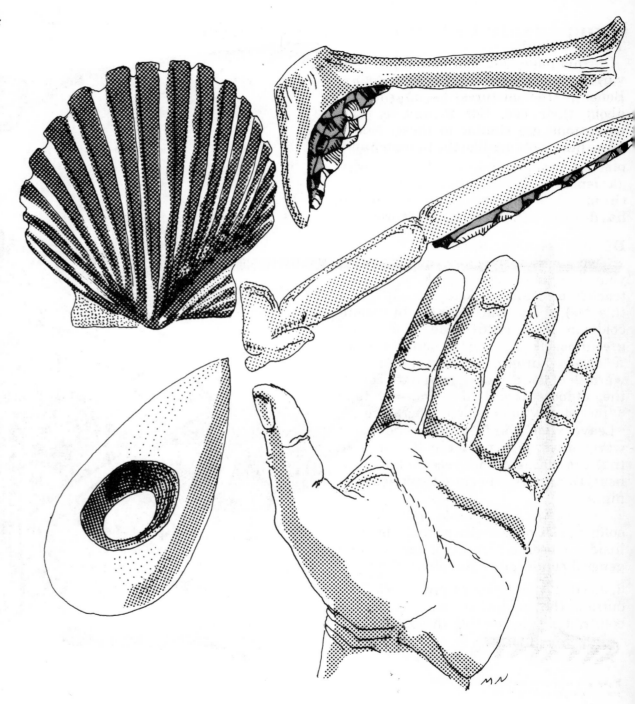

# Today's Tools

On this page, you will find pictures of various tools used in gardens today. Beneath the pictures is information about their use. See if any of your inventions are similar to these. Notice how these tools are like the human hand.

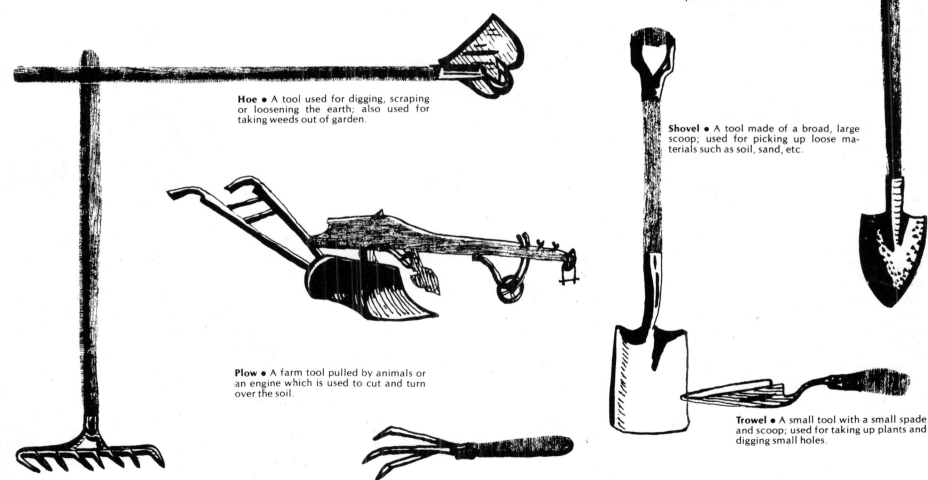

**Pitchfork** • A large long-handled fork used for lifting and tossing compost, hay, etc.

**Spade** • A tool used for digging and cutting the ground. Spades have a pointed edge for digging and a flat edge which is pushed with the foot.

**Hoe** • A tool used for digging, scraping or loosening the earth; also used for taking weeds out of garden.

**Shovel** • A tool made of a broad, large scoop; used for picking up loose materials such as soil, sand, etc.

**Plow** • A farm tool pulled by animals or an engine which is used to cut and turn over the soil.

**Trowel** • A small tool with a small spade and scoop; used for taking up plants and digging small holes.

**Rake** • A tool with metal or wood teeth for scraping or pulling things together.

**Claw** • A small tool used for scraping.

# Companion Planting 25

Long ago, gardeners planned their gardens according to which plants 'liked' one another. They found that some plants would grow better if planted next to certain other plants. This is called **companion planting**. As cities grew large, and people needed to grow more and more food, farmers stopped companion planting and lost its benefits.

Today, people are remembering the wisdom of their ancestors. More and more gardeners are planting 'companion gardens'. Why do they work? People who are doing research say that:
- Some plants with deep roots help plants with shallow roots. The deep roots can increase the amount of minerals brought to the surface of the soil for the shallow roots.
- Some plants attract helpful insects.
- Some plants enrich the soil around them with nutrients.
- Some plants such as mint or garlic contain chemicals which repel insects.

Plant your garden by the companion method. Be sure to plant for 'dislikes' as well as 'likes'.

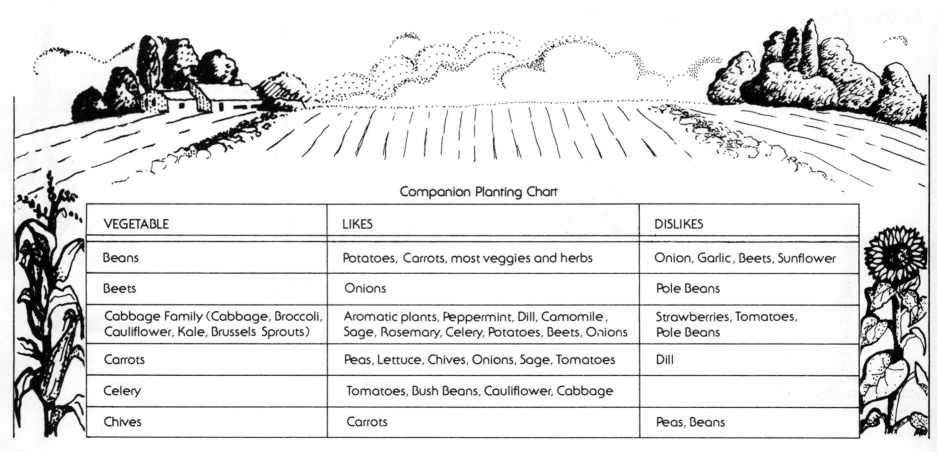

Companion Planting Chart

| VEGETABLE | LIKES | DISLIKES |
|---|---|---|
| Beans | Potatoes, Carrots, most veggies and herbs | Onion, Garlic, Beets, Sunflower |
| Beets | Onions | Pole Beans |
| Cabbage Family (Cabbage, Broccoli, Cauliflower, Kale, Brussels Sprouts) | Aromatic plants, Peppermint, Dill, Camomile, Sage, Rosemary, Celery, Potatoes, Beets, Onions | Strawberries, Tomatoes, Pole Beans |
| Carrots | Peas, Lettuce, Chives, Onions, Sage, Tomatoes | Dill |
| Celery | Tomatoes, Bush Beans, Cauliflower, Cabbage | |
| Chives | Carrots | Peas, Beans |

| VEGETABLE | LIKES | DISLIKES |
| --- | --- | --- |
| Corn | Potatoes, Peas, Beans, Cucumbers, Pumpkin, Squash | |
| Cucumbers | Beans, Corn, Peas, Radishes, Sunflowers | Potatoes, Aromatic Herbs |
| Tomato | Chives, Onion, Parsley, Asparagus, Marigold, Nasturtium, Carrot | Potato, Fennel, Cabbage |
| Eggplant | Beans | |
| Peas | Carrots, Turnips, Radishes, Corn, Cucumbers, Beans, most veggies and herbs | Onions, Garlic, Potato |
| Squash | Nasturtium, Corn | |
| Onions (& Garlic) | Beets, Strawberries, Tomato, Lettuce, Some Camomile | Peas, Beans |
| Lettuce | Carrots & Radishes (along with Lettuce make great growing team), Strawberries, Cucumbers | |
| Radish | Peas, Nasturtium, Lettuce, Cucumbers | |
| Parsley | Tomato, Asparagus | |
| Potato | Beans, Corn, Cabbage, Horseradish (plant at corners of patch), Eggplant, Marigold (as lure for Colorado potato beetle) | Pumpkin, Squash, Cucumber, Sunflower, Tomato, Raspberry |
| Pumpkin | Corn | Potato |
| Spinach | Strawberries | |
| Sunflower | Cucumbers | Potato |
| Turnip | Peas | |

# Drawing Your Garden Plan                26

Remember the seeds your class decided to plant? In this activity you are going to make the final plan for your entire garden. Use your journal or a large sheet of paper for the plan. Record all important information about each vegetable in your journal.

## DOING
1. Measure your garden space. Record this on your plan.

2. Use a compass to mark on the garden site and on your plan the directions east, west, south, and north. Make a directional indicator using sticks or stones.
3. There are several good ways to lay out your garden. One is to plant each type of vegetable in rows across your entire garden space. See Sample Plan #1 on the next page.

Another way is to make what are called raised beds. See Sample Plan #2. Beds are made by raising the soil level in certain areas. In the sample plan, each bed is 8 feet long, 4 feet wide, and 6 to 8 inches high. Raised beds are a good way of planting because the higher soil level provides good water drainage and **aeration**.

Leave about 3 feet of walking space between each bed. Inside each bed you can plant several rows of vegetables.
4. Make your rows or beds run from east to west if you can. Also make sure that your tallest plants (corn, for example) are on the north so they won't shade the smaller plants.
5. Make a list of all the vegetables and flowers you elected on page 22.
6. Check the companion planting section to see which of the vegetables and flowers you've chosen like to be near each other. Decide which ones will be companions. If you are using raised beds, companion plants can be put together in the same bed.

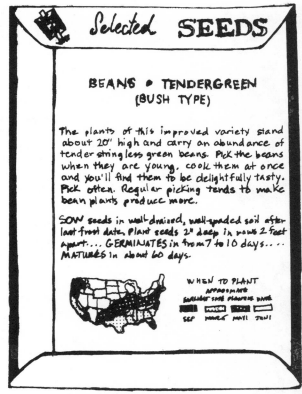

7. Decide how many rows you want of each vegetable.
8. Check the seed packages for the space each vegetable needs between rows.
9. Record the date you will put the seeds or seedlings in the ground.
10. Draw your final plan on the front of this card. Remember:
A. East-west direction of rows or beds.
B. Which plants need sun and which need shade.
C. Which vegetable is going in which row or bed.
D. Space between rows.
E. Companion planting.
F. Date seed or seedling is planted.

# Sample Garden Plans     27

## Plan # 1

ROW GARDEN    —NORTH—

| CORN | CUKES | COMPOST |
| CORN | PEAS (THEN CABBAGE) | |
| RADISH (THEN CORN) | POTATOES | |
| LETTUCE | BEANS | |
| CARROTS | TOMATOES | |
| SPINACH (THEN BEETS) | TOMATOES | |
| SQUASH | ONIONS | |
| SQUASH | PEPPERS | |

## Plan # 2

RAISED BEDS    —NORTH—

| SQUASH / MELONS | CORN / CORN | BEANS / CUCUMBER ON TRELLIS |
| | BEETS / CABBAGE | POTATOES / MARIGOLDS |
| | LETTUCE / CARROTS | PEAS (THEN KALE) / RADISHES (THEN ONIONS) |
| COMPOST | TOMATOES / HERBS* | TOMATOES / NASTURTIUMS |

\* SAVE SPACE HERE TO PLANT SOME HERBS LIKE: CHIVE AND PARSLEY (MARCH); DILL (APRIL) OR BASIL (MAY).

NOTE: COMPANION FLOWERS AND LEAFY GREENS MAY BE SPACED ALONG PATHS.

| | |
|---|---|
| Beans | Don't put in cold, wet soil |
| Beets | Plant seeds as soon as ground can be worked |
| Broccoli | Begin seed germination indoors, then transplant. |
| Cabbage | Start indoors, plant seedlings firm & deep |
| Carrots | Soil must be fine and loose — take out all rocks |
| Cauliflower | Start indoors. A challenge— needs plenty of water |
| Collards | Easy to grow |
| Corn | Needs lots of space |
| Eggplants | Start indoors — plant only when warm |
| Lettuce | Shade lover — grow plants close together |
| Melons | Start seeds indoors |
| Onions | Grown from sets (tiny bulbs) |
| Peas | Plant early — pick pods as soon as they ripen |
| Peppers | Start seeds indoors |
| Potatoes | Soil free of lumps, potatoes in pieces w/"eyes" — place in 6" deep trench. |
| Pumpkins | Takes space but it's worth it at Halloween. Plant on edge of garden. |
| Radishes | A fast crop and easy to grow. |
| Summer squash | Plant in small mounds |
| Tomatoes | Start indoors 6-8 weeks before putting out. Stake plants when outdoors |

# Planting the Garden 28

## DOING

You will need:
- shovel
- rake
- string
- pegs or sticks

1. Turn over the soil in the garden and break up any large clumps. Take out any clumps of grass and large and medium stones. Don't try to remove small pebbles; these help water drainage.
2. Spread compost or another type of organic fertilizer on the garden. Use a shovel to put it on, then spread it evenly over the garden with a rake.
3. Let the garden rest for at least a day. This will allow the fertilizer a chance to seep into the ground.
4. Prepare any beds. Make the beds by raising the soil level 6 to 8 inches higher than your paths. Mark off your beds with string and sticks.
5. Follow your garden plan, mark each row you will plant with a string and two sticks. Place a stick firmly in the ground at one end of the row. Tie the string to it. Have a partner pull the string along in a straight line to the other end of the row. Place another stick in the ground and tie the string to the stick.
6. Mark each row with the name of the vegetable or flower you will plant there and the date on which you will plant it.
7. Soak your seeds in water for 24 hours before they are to be planted. This will help them germinate.
8. Read the seed package to find out how far apart to plant each seed. Plant your seeds directly under the strings connecting the stakes. Use your fingers or a pencil to make holes for the seeds. It is important to plant the seeds at the right depth. Follow the instructions on the package or plant them four times as deep as they are big. Pat the soil down over them when you are done. For small seeds like carrots, it is best to just sprinkle them over the soil and pat them down.
9. Follow any special instructions (like making a trellis or teepee for peas or beans).
10. Gently water the soil where you planted seeds. Don't let the soil dry out, especially when seeds are germinating. Be careful not to over water.

1.  2.  4.  5.  6.  7.  8.  9.

# 29
## Planting Seedlings

When planting seedlings, above all be gentle. Remember that you are uprooting a living thing and expecting it to make a new home and grow.

Here are the steps to follow:

A. Do your transplanting on a cloudy or overcast day or in the late afternoon. A hot sun can wilt a seedling.
B. Water the seedling 1 hour before planting.
C. Dig a hole that is twice as big around as the root mass and as deep as the roots of the seedling. Wet the hole. Add a little compost.
D. Remove the plant from its container, either by turning the container over and tapping it on the bottom or by digging carefully around the edge with a spoon handle.
E. Place the seedling in the hole and gently fill around it with soil, tamping down firmly after each fill. Remember to be gentle.
F. Be sure all the roots are covered. Be sure not to cover any leaves or the growing tip of the stem.
G. Make sure that your rows and the plants within each row are far enough apart. Check the seed packages for instructions.
H. If you have tomatoes or peppers, make little paper collars to surround the plants. These are to keep away the cut worms that eat tender young stems. Plant some with and some without paper collars and observe whether they work.
I. Watering is especially important until the roots can grow deeper. Water just after transplanting and each time the soil becomes dried out. Wilted stems and drooping leaves are signs that the plants need water.

A special growth-booster is a bucket of water with manure or compost mixed in. Let the mixture sit over-night and then give your plants a treat!

Your garden is all planted! The fun has just begun! We still have to observe the way the environment interacts with vegetables planted outdoors, and that is why we must learn about weeds, insects, flowers, and the final return of the plant to seed.

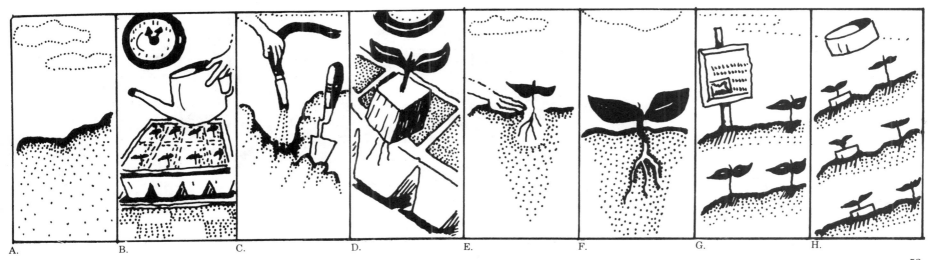

A.   B.   C.   D.   E.   F.   G.   H.

53

## 30

# Scarecrows

While you're working in your garden, the birds won't come near it. But once you've planted seeds and gone away, some birds may fly into the garden to eat the seeds.

To keep the birds away during seeding time, put a scarecrow in your garden. A scarecrow is something that looks enough like a person to fool the birds into thinking it is a person. If you design one with flappy arms or a hat that moves in the wind, it will do an especially good job of keeping the birds away. You can also hang some shiny metal strips on it that will glitter and bang together. The scarecrow will be your friend, so have fun making it.

You can move the scarecrow out of the garden after all your seedlings have sprouted their first leaves. Now when birds come there are no seeds for them to eat. Instead, they eat insects that would otherwise eat your plants.

# Community Gardens

## 31

### Growing Food with Your Neighbors

What do you do if your back yard is too small or shady for growing a garden? Or if you live in a city apartment and don't have any land at all? Or if you would like to garden with other people?

Many people who face these problems but still want to garden are turning vacant city land into **community gardens.** Community gardens bring people together to share land, tools, friendship, and new ways of growing food. Community gardens change unused land into a food-growing and money-saving **resource** for neighbors.

Community gardens are not a new invention. They are as old as the science of farming. For ages, people have come together to share land and help each other survive.

In Boston, Julia Brown is the coordinator of a 20-plot community garden. In 1978 Julia grew nearly $700 worth of vegetables on her 20 by 20 foot plot. Julia grew up in the **rural** south of the United States and then moved to Boston. In her garden she grows several vegetables that are rarely grown in Boston but are commonly grown in the South. They are **collards, peanuts, okra,** and **mustard greens.**

Community gardens are also great places to learn about vegetables eaten by people from other countries. Below is a list of 3 **ethnic** groups and some of the different foods they are growing in their community gardens.

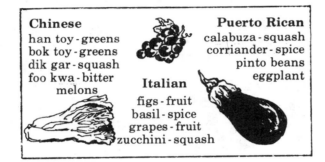

**Chinese**
han toy - greens
bok toy - greens
dik gar - squash
foo kwa - bitter melons

**Italian**
figs - fruit
basil - spice
grapes - fruit
zucchini - squash

**Puerto Rican**
calabuza - squash
corriander - spice
pinto beans
eggplant

### DOING

1. Find out more about the history of community gardens. Key words or topics to look for in your research are: New England Commons (colonial agriculture); collective farms (farms owned by workers); Victory Gardens (World War I and World War II); ancient farming (Egypt, Greece).

2. Write down each of the ethnic foods listed on this page that you are not familiar with. Find out more information about each of them in a seed catalogue or from your library. Can you grow any of these foods in your climate?

3. Below are ten vegetables that are listed in the Sample Garden Plan #2 on p. 51. Calculate how much money you could save if you grew the following amounts in your garden. To do this activity you will need to find out the current price of each vegetable.

| | |
|---|---|
| 20 lbs. yellow squash | 10 lbs. beans |
| 10 heads lettuce | 24 cucumbers |
| 10 lbs. carrots | 30 lbs. potatoes |
| 10 lbs. tomatoes | 5 lbs. peas |
| 20 ears corn | 5 lbs. radishes |

# Old MacDonald Had a Farm

## Disappearing Farmland

At one time most of the food people ate came from their own yards or from small local farms. The food grown was high in quality and freshness and low in cost. The farms were usually run by a family who sold their vegetables and dairy products to local markets. Food was bought, cooked, and stored as needed.

Today things have changed. Most of the food Americans now eat comes from large out-of-state farms or even food factories! Go to any supermarket and look around the aisles. Almost all of the food on the shelves is either canned, packaged, frozen, or **processed** in some other way. These processed foods are made by large food companies who own large farms and sell their products around the world.

Because many of these "new" foods save cooking time and are attractively packaged, people are buying them instead of fresh foods. The effects of this change on the food buyer, small farmer, and food companies are compared below.

| Modern food buyer | Small farmer | Food companies |
|---|---|---|
| saves time less fresh may be less nutritious more expensive | many going out of business | large sales and profits |

Since 1935 over 4 million small farms have gone out of business because they could not compete with large food companies. In Massachusetts over 200 farms

THE WAY IT WAS:

were going out of business each year during the 1960's and 1970's.

This loss of valuable farmland aroused many people's interest in helping the small farmers in their state. One solution was for farmers to sell their food directly to city dwellers in what is called a **Farmer's Market.** At one of these markets farmers set up fruit and vegetable stands in the streets so people can walk along and choose the best **produce** they can find. A Farmer's Market not only helps the farmer but also provides shoppers with fresh, nutritious, and low cost products. In 1980 there were over fifty Farmer's Markets that operated weekly in Massachusetts.

Each day more people are realizing that fresh foods are better than processed foods. By choosing fresh foods more often, people help local farmers to stay in business.

TODAY'S FOOD SYSTEM:

Old MacDonald *had* a farm—and soon he may be able to buy it back. Try to buy locally grown foods—you'll not only help the farmer but get a better product in return.

## DOING

1. Below are some of the price comparisons between a Farmer's Market and a supermarket. Figure out how much money you would save if you bought one pound of each food listed.

|  | Average local store price (1978) | Farmer's Market price (1978) |
|---|---|---|
| Boston lettuce | .72/lb. | .50/lb. |
| Romaine lettuce | .36/lb. | .14/lb. |
| Iceberg lettuce | .79/lb. | .50/lb. |
| Broccoli | .77/lb. | .53/lb. |
| Spinach | 1.32/lb. | .38/lb. |
| Cucumbers | .62/lb. | .29/lb. |
| Beets | .53/lb. | .40/lb. |

2. Write your local state agricultural agency to find out if there are any Farmer's Markets in your state, or about their efforts to promote local farming.

3. Foods are processed in many ways. Processing can be as simple as cooking and freezing a food or as complex as adding dozens of chemicals to color, flavor and preserve a "new" product. Processed foods usually have long lists of ingredients on their labels. Go to your kitchen and look for the *most* processed foods. Make a list of them and their ingredients. Research the ingredients to find out if any of them may be harmful to your health.

# Roots • Waterworks of Plants

## 33

While we are waiting for our seedlings to mature, let's look at how a plant grows from a tiny seed into a large plant that may produce millions of seeds.

The first things we'll look at are roots. Here are some of the ways the word **root** is used in our language:

money is the *root* of all evil
*rooted* in the spot
*rooting* around
putting down *roots*
*root* beer

Think about the word. What does it mean to you?

Roots perform three important jobs for a plant.
1. They anchor the plant in the soil. Did you ever try to pull up a tree, or a bush, or even a dandelion? Difficult isn't it? That's because the roots are holding the plant in the ground.
2. Roots absorb water and minerals from the soil and transport them to all parts of the plant. A mature apple tree absorbs 50 gallons of water a day!
3. Roots store food for the plant's future needs. We use the food which some vegetables have stored in their roots when we eat carrots, or beets.

## DOING
### Root-Viewing Box

To see how roots work and how different kinds develop, we are going to use some root-viewing boxes. Since we cannot see through the soil in our garden, we will use these boxes to see the underground action of roots. We will plant different types of seeds in the boxes to see different types of roots. By the time your plants are growing in the garden, you will understand what is going on underground.

You will need:
- root boxes—see illustration
- soil mixture
- seeds (radish, carrot, lettuce, etc.)

1. Add the soil mixture to the root boxes.
2. Plant several types of seeds in the boxes. Read the instructions on the seed packages for planting depth. Mark and label where you planted each seed.
3. Water the seeds.
4. When the boxes are not being viewed, cover the front window. Roots like darkness.
5. Watch the seed as it begins to germinate. Which part sprouts first, the root or the leaves? Take a guess why based on what you know about the needs of plants. How does a root know to grow downward?
6. Look very closely at your roots. Do you notice any very thin, hair-like threads on the larger roots? These are called root hairs; a grown plant can have miles of them. In fact, a certain type of grass was measured to have 6,600 miles of root hairs coming from one plant!

Root hairs are the real wonder workers of the system. They absorb all the waters and minerals for the plant. The water that most plants take in through their root hairs is not from puddles and streams. It comes from a thin coating of water that is around each grain of soil. The root hairs absorb this water into the plant. Each tiny drop of water that is absorbed has mineral nutrients from the soil in it.

As the root hairs grow, they find new supplies of water and nutrients. This supply of water and soil nutrients is moved upwards from the root hairs. First it goes through pipelines inside the larger roots, then into the stem, and finally into the leaves. What do you think happens there?

7. What type of root system does each plant have? The two types are:

A. **Tap Root** - This root system is thick and long. It has one main root with smaller roots growing from the sides. Tap roots are so powerful that they can break through rocks and concrete with ease.

B. **Fibrous Roots** - These are thin and there are many of them growing down from the ground level. They are also powerful. Grass plants have fibrous roots that can grow in the cracks in sidewalks. Look for them on your way home from school.

# Stems • Transporters & Supporters

Can you imagine a stem so large that 100 men on horses could hide behind it? There really was a stem like that, the famous 60-foot wide chestnut tree on Mount Aetna in Italy. In your classroom, measure how wide this tree was.

Stems have many sizes, shapes, and uses. Have you ever heard about the giant redwoods in California? Their stems reach more than 300 feet straight up into the sky. Dandelions have stems which are hollow inside. Celery and rhubarb are stems that people eat. All the wood in your house and the paper in this book came from stems of trees.

Have you ever looked inside a stem? When you do, you will see many small tubes that connect the root pipelines with the tiniest veins in the leaves. These tubes transport water, minerals, and other nutrients to all parts of the plant. This mixture of water and nutrients is called **sap**.

Besides being transporters, stems are also supporters. They hold the leaves up into the air and sunlight. Why is this important?

# DOING
## The Power of Stems

To see a stem's transport tubes carry water, try this easy experiment.

You will need:
- a clear glass or plastic cup
- 1 stalk of celery with its leaves
- red or orange food coloring
- water
- magnifying glass

1. Fill the glass ½ full of water and add 5 or 6 drops of the food coloring to it. Mix.
2. Put the celery in the colored water.
3. Place it in a sunny window.
4. After 1 day, look at the veins in the leaves to see if the colored water has been transported there.
5. After 2 days, break open the stem. Look for the transport tubes, which will be colored red. Use your magnifying glass to look at the leaves.
6. Draw a picture of the path of water through the celery.
7. Draw the pattern of the veins in the leaves.
8. At home or in your neighborhood, collect 4 examples of stems. Don't forget vegetables in your search. For example, part of broccoli is a stem. Mount your stems on paper and label them. Notice how many different kinds of stems there are. Some are hard, some are soft, some are clear, some are hollow, and some are filled.
9. Look at the rings in an old tree stump. Look closely using a magnifying glass. You will see that these rings were actually the transport tubes for the tree.

# The Water Cycle

## 35

In nature, water is used and reused in an endless cycle.

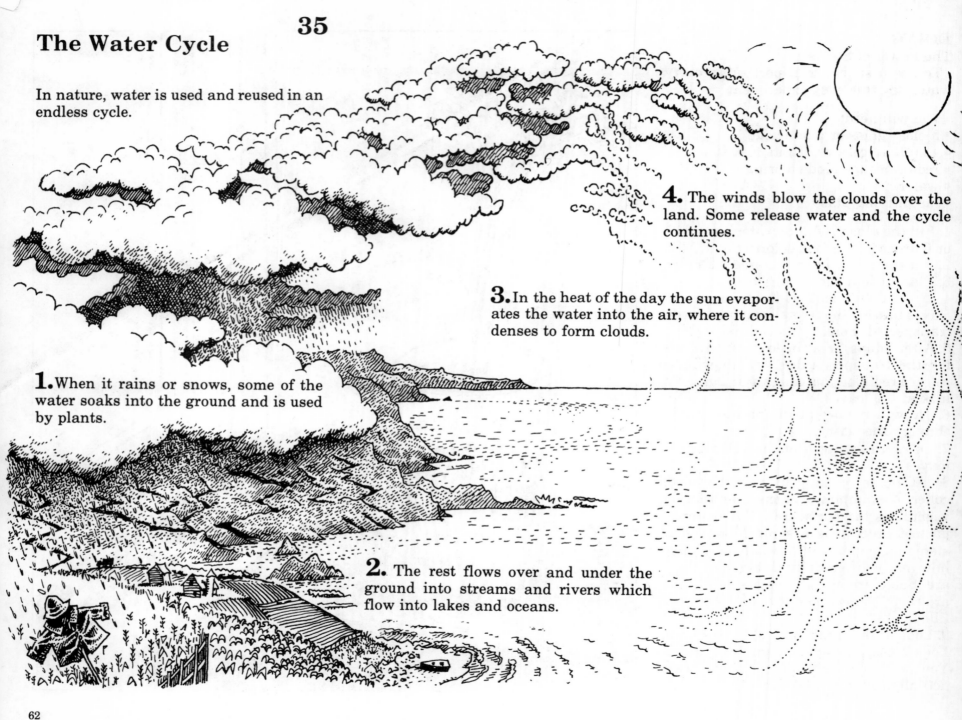

**1.** When it rains or snows, some of the water soaks into the ground and is used by plants.

**2.** The rest flows over and under the ground into streams and rivers which flow into lakes and oceans.

**3.** In the heat of the day the sun evaporates the water into the air, where it condenses to form clouds.

**4.** The winds blow the clouds over the land. Some release water and the cycle continues.

The amount of rain that falls is important in any climate. When the rain did not fall, the Navajo Indians used the sign shown above in their rain dances. It represents a thunderstorm. They knew how important rainwater was for the growth of their food. Without water a plant will dry up and die. With too much water it will drown.

Always remember that your garden will need water to grow, but that too much water will do it harm.

**Tips on Watering**

1. Plants should be watered every few days to once a week depending on weather and soil conditions. To find out if your garden needs water, dig down between the rows a few inches. If the soil is dry, then your garden needs to be watered.

2. The best time to water is in the morning. Avoid watering during the hottest part of the day.

3. Material such as hay, leaves or woodchips can be added to the soil around your plants. This method is called mulching. Mulching the soil conserves water and prevents weed growth.

# 36

# The Tree Experiment

About 350 years ago, a man named Jan Van Helmont decided to find out how plants grow. At that time, most people thought plants ate soil. Jan wasn't sure this was true, so he set up an experiment to find out for himself. He planted a small, 5-pound willow tree in a pot of dry soil weighing 200 pounds. Jan figured that if the tree ate the soil, then the weight of the soil should get less and less.

For five years Jan watered and took care of the willow. It grew very well and became a handsome 169-pound tree. Then Jan weighed the soil again. He was careful to let the soil dry out so that it would be as dry as when he first planted the tree. The soil tipped the scales at 199 pounds and 14 ounces, only 2 ounces lighter than the original 200 pounds! Where did the tree get the food to grow 164 pounds? Jan thought it all came from the water he added. Where do you think it came from?

# Leaves • Food Factories

For many years, people were bothered by the question, "How does a plant get the food it needs for growth?" People knew that they themselves needed to eat to survive. "But," they asked, "does a plant eat or does it make its own food?"

After a lot of hard thinking and investigating, the mystery was finally solved. It was discovered that the leaves of every vegetable, tree, and blade of grass contain a kind of 'food factory'. Although a leaf may appear as thin as paper, it really has an inside which makes food from **light**, **air**, **water**, and **soil**! The leaf is like a sandwich, with all the equipment for making food packed between its upper and lower surfaces (see illustration on page 68).

One scientist found that a large tree working for 10 hours a day from May to September could create 3,600 pounds of food. A garden vegetable such as a watermelon makes about 20 pounds of food, while a carrot plant makes about 4 ounces of food. Just imagine how much food is made each year by all the plants in the world!

The seedlings you planted have this 'food factory' working for them right now. The food they create will be used for growing and for producing the mature plant.

## How the 'Food Factory' Makes Food

The 'food factory' goes into operation as soon as sunlight strikes and penetrates the surface of a leaf. Once inside the leaf sandwich, the sunlight is trapped. The trapped **sunlight** then combines with **air** and **water** to make **sugar**.

This combining of sunlight, air, and water to make sugar is called **photosynthesis**. Photosynthesis is the first step in the making of all food on earth. It happens everywhere that sunlight falls upon green plants.

Rows of Cabbage Leaves

## Summary of Photosynthesis

> Green plants combine
> Sunlight + Air + Water = Sugar

The next step in food making comes when sugar is:

> turned into starch
> **or**
> combined with the soil minerals and nutrients

These two combinations make the building blocks which form new leaves, larger stems, bigger roots, fruits, and seeds.

Can you name these building blocks? (We talked about them on pages 13 & 14.)

## Summary of Food Making

> Green plants combine
> Sunlight + Air + Water + Soil
>       = All Food

### Where do the materials necessary for photosynthesis come from?

**Sunlight** - The sun is a blazing yellow star one million times bigger than the earth. The light and heat from the surface of the sun take about 8 minutes to reach the plants on earth. Look at your seedlings and other plants in your classroom. Are they bent toward the sunlight? Why?

fruit of corn

**Air** - This invisible material completely surrounds the earth. It is everywhere. Even though you can't see it, you can feel it when the wind blows. Just as soil is made up of many parts, so too is air. Two of the main parts are **carbon dioxide** and **oxygen**. Plants use the carbon dioxide in photosynthesis. Animals and people use the oxygen to breathe.

**Water** - As you already know, water comes from rain which soaks into the soil. Roots absorb the water and send it through the root and stem transport tubes to the leaves.

**Soil Minerals and Nutrients** - As you remember, these materials come from the breakdown of rocks and the decay of dead plants and animals. They are absorbed with water from the soil. Once inside the roots, these materials are transported to the leaves.

### A Secret of Life

This secret is very exciting, for it tells us the reason why we are alive! When plants make sugar by photosynthesis, they release oxygen. Take a deep breath. You just inhaled a lot of oxygen that was made by plants. In fact, all the oxygen in the air came from plants!

Breathe out. You just released carbon dioxide from your body. Plants use this carbon dioxide to make food for themselves and us!

### The Secret Revealed

> Green plants combine
> Sunlight + Air (carbon dioxide) + Water + Soil = All Food + All Oxygen

# A Tour of the Factory 38

Look at the picture of the leaf on this page. We magnified the leaf sandwich to show its equipment for making food and oxygen.

**Topside** - The roof of the 'factory' is covered with a clear coating of wax to protect it. Sunlight passes through this coating to reach the site of food production.

**Palisade Cells** - These long sausage-like cells are underneath the top side of the leaf. They trap the sunlight that reaches them. The palisade cells are the main site of food production.

**Chlorophyll** - Inside the palisade cells are many small grains. These grains contain green chlorophyll. The chlorophyll grains are microscopic, but there are so many of them thay they alone give leaves their green color. Chlorophyll operates the 'factory'. It is responsible for combining sunlight, air, and water to make sugar. Without chlorophyll, a plant cannot make food.

**Spongy Cells** - These cells have many different shapes. They contain a few specks of chlorophyll which trap any remaining light rays for photosynthesis.

**Veins** - Some of these open tubes carry water and nutrients to the chlorophyll grains for photosynthesis, and some carry food made by photosynthesis to other parts of the plant. If we could enter one of these veins and follow its passageway, where would we come out? Veins can easily be seen without a microscope. Their beautiful patterns reach every part of the leaf.

**Underside** - The underside of the leaf also has a protective wax coating. On this side are openings to the inside of the leaf. These openings are called **stomata**. The word comes from a Greek word meaning "mouth". Stomata look a bit like the lips of a circus clown. The stomata let air into the leaf for food production. They also release water and oxygen out of the leaf. Stomata are so tiny that one inch of leaf surface may contain 250,000 of them.

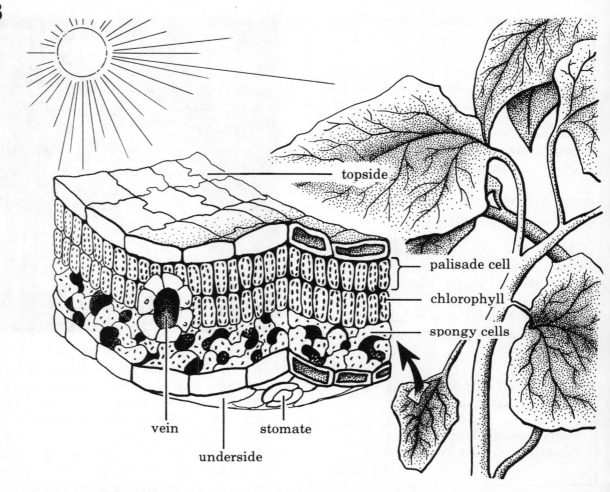

# Adaptations • How Plants Survive

## 39

In order to make food, plants need sunlight, air, and water. Too much or too little water or sun can kill a plant. To help control the amount of sun and water they get, some plants have developed amazing features. Here are a few:

**Bean Plants** will tilt their leaves downward on a hot day so that the sun cannot shine directly on them. This cools the surface and reduces water loss.

**Carrots** have delicate, lacy leaves so all the little parts can spread out in the sun. If they had one solid leaf which covered the same area, it would demand more water than the roots could supply.

**Tomatoes** have soft hairs on all the stems and leaves. These hairs slow down the breezes and shade the stem. The slower the air moves across the leaves, the less water it can pick up and carry away.

**Cabbages** have thick, waxy leaves and an enlarged stem which act as reservoirs to hold extra water in case of dry days.

**Grasses** have narrow leaves so they can live in open fields with sun all day. Smaller leaf surface reduces water loss.

Notice how plants which grow in shady places have wider, darker leaves to absorb more light. The darker the leaf, the more light energy it can absorb.

Take this list outside with you and look for examples of the following:
- Leaves tilting or folding away from the sun
- Lacy, cut, or frilly leaves
- Hairs on leaf or stem surface
- Thickened stems for reservoirs
- Thin leaves for life in the open
- Large leaves for shady places

# Weeds • Good Plants in Wrong Places

Have you ever thought about what a weed is? Did you ever think that a weed could be a tomato, or a carrot, or even a beautiful bush of roses? It's true! A weed is simply something that you have decided you don't want growing in your garden. That means that if you have a flower garden, a vegetable (perfectly good for eating) would be considered a weed and taken out of the garden. In this situation, the vegetable would compete with the flowers for valuable water, nutrients, and sunlight. The vegetable's roots would crowd the flower's roots. The vegetable's leaves would shade the flower's leaves.

Think about some of the plants we consider weeds: dandelion, goldenrod, ragweed. Did you know that most of these "weeds" were considered sacred plants by Indians and early settlers? They were used to heal people, to preserve strength, and to eat in delicious, vitamin-filled salads.

On the next page are some pictures of common weeds that are likely to grow in in your garden. Before you throw them away, why not make some dandelion tea, or a lamb's quarter salad? Your class might want to leave a little patch of garden space to grow wild - just to see what kind of "weeds" will grow.

### DOING

1. Make a weed collection from your garden. Be sure to pull them up roots and all. If you don't, they will grow back quickly. Instead of throwing the weeds away mount them on paper. Take a weed walk. Collect leaves, flowers, stems, and roots. Then go to the library and find a book on weeds or herbs and see if you can match what you have with what's in the books. Make a display of the different types telling their history. Put the ones you don't use in your compost pile so that their bodies can enrich the soil.

2. Make a weed cookbook. Make sure you have identified your weeds as ones that can be eaten, and then find out how to prepare them. Write down your recipes and your comments about how they taste. **Remember not to eat any weed until you know it's safe!**

3. Look at any weed in the garden and give it a name. Make up a story about what you think it can do. This is how many good things about plants were discovered — by people using their eyes, ears, noses, and imaginations. Try using yours!

**Burdock** - The Japanese boil the root of burdock to eat as a vegetable. The young leaves and stalks are also eaten.

**Clover** - Eat the flower heads of clover in a salad or make a tea by pouring hot water over the flowers and letting the tea steep (sit) for ½ hour. The tea is tasty with honey. The soft clover flowers feel good on the feet. Try running through a field of it sometime. Clover feels so good that Swiss mountaineers, who wore hard wooden shoes, used to stuff the clover blossoms inside to make them more comfortable.

**Dandelion** - This name in French is *dents-de-lion*, or teeth of the lion. Look at the plant carefully to see why this name might have thought of. Do you see any part that looks like a lion's tooth? Every part of the dandelion can be used for food. The flowers are good for making tea and wine. The leaves, when they are young and tender, make a salad that is very good for you. The roots are often dried and made into a coffee-like drink.

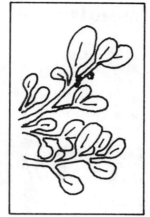

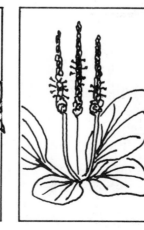

**Purslane** - This is a commonly stepped on, low-growing plant, whose juicy leaves are good for nibbling or cooking in soup. You may as well like purslane, as it is almost impossible to get rid of. Even after you've pulled it up, there is enough energy in the thick leaves for the plant to make tiny seeds in a few hours!

**Lamb's Quarter (Pigweed)** - The soft leaves of this plant make some of the best salad greens. Eat them raw or cooked. They are delicious. They will probably grow with your first lettuce plants and continue to grow all season. Watch for them. You won't be disappointed.

**Plantain** - Everybody tries to get plantain off their lawns in the summer, but they probably wouldn't try so hard if they knew what it can do. Like many other "useless" weeds, plantain makes a great salad. It also takes away the pain of insect bites. Crush a plantain leaf and rub it on the bite... the pain will be gone in no time.

# Garden Insects • Pests & Friends

Before you spray your garden with Raid or launch an all-out attack on everything that crawls or flies, wait a minute.

Put down all those chemical poisons that do everybody harm. It's likely that you don't need them at all. Did you know that most insects are helpful to people? Here are some of the ways we depend on them:

- They help most plants produce fruit and seeds. Bees, butterflies, and a thousand other insects are needed for this important job.
- They turn over and plow the soil. Their underground tunnels allow air and water to reach the plant's roots. Beetles and ants are good examples of this.
- They become food for fish, birds, and other animals.

Just imagine what the world would be like if all the insects were killed. Insects are real friends.

Some insects are garden pests because they like to eat the plants you want to grow. When people use chemical poisons, they not only kill the pests but many of our friends, too. Many times the poison doesn't even work. Chemical poisons can also be dangerous to the soil, to you, and to other living things.

There are many other ways to keep your garden pest-free and safe. This year try to use:

**Repellent Plants** — Many pests don't like certain types of plants. Marigold and nasturtium flowers are examples. Plant them around your garden.

**Safe Chemicals** — These chemicals are made from plants. Their names are **pyrethrum** and **ryania**.

**Insect Friends** — Many insects like to eat insect pests. Ladybugs and the praying mantis are examples. One ladybug can eat 500 aphid pests in one season. Many companies sell ladybugs and mantises. Why not try them out and watch them do their job?

**Homemade Repellent** — Try this old-time natural recipe for pest control.

1. Chop together:

   3 hot peppers
   3 onions
   1 clove of garlic

2. Mix the above in 1 gallon of water and let the mixture soak overnight.

3. Transfer the mixture to a spray bottle and spray on plants when needed.

### DOING

1. Go out and find the different insects, caterpillars, and worms that live in or visit your garden. What kinds of food do they seem to like? Are they helpful or harmful or just neutral? Draw pictures of them.

2. Set up experiments using the different types of safe pest controls. Record your observations in your journal. Use a repellent only if you are sure the insects are hurting the plants.

# Insect Control

It is important to know which insects are helpful to our plants and which are harmful. If young plants are bitten at the roots, leaves show holes, parts of plants suddenly wilt, or a choice vegetable is nibbled to bits, then probably the insects have found your garden. Do not reach for a chemical poison. Find out which insects are doing the damage and spend some time learning about their lives so you can control them.

Careful observation should show that:
1. Certain insects do certain types of damage and can be controlled without poisons.
2. The plants under attack tend to be weaker ones. Perhaps they were planted too close together, or they were just transplanted, or they have not been kept in soil rich in compost.
3. If you use a poison to kill the pests, you are likely to kill its predator as well. A **predator** is an animal that lives by eating other animals. If you kill the predators, new insect pests will appear and you will have won nothing.

Here is some information about the most common garden insects.

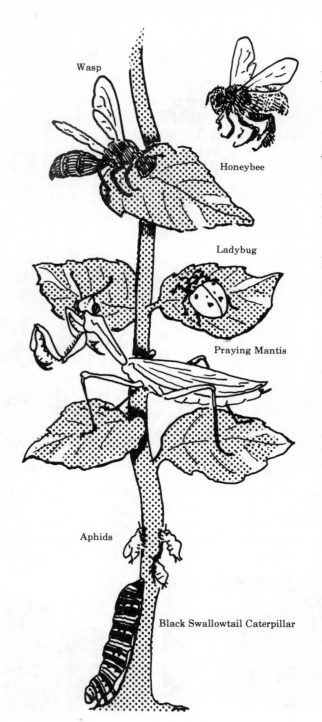

## Helpful Insects

**Honey Bees**: Bees spread pollen, which fertilizes plants so they can produce fruit and seeds. In the garden, bees are busy and will not sting you unless you bother them.

**Wasps**: The police of the garden, they spend all day catching leaf-eating worms. Be careful; they may sting you.

**Ladybugs**: They are terrific aphid eaters. The baby ladybugs eat even more aphids than the adults do.

**Praying Mantises**: Largest of the insect predators, they catch other insects in their front legs and eat them whole.

## Harmful Insects and How to Control Them

**Aphids**: Tiny but mighty, they can weaken a plant by sucking its juices. Spray them with a solution of strong herbs, garlic, onion, or soap (not detergent).

**Black Swallowtail Caterpillars**: They will eat your carrot, parsley, or dill leaves. You can pick them off and throw them to the birds. Get the jump on them by checking for any egg clusters underneath leaves.

None of the smooth-skinned caterpillars will hurt you. If you want to watch black swallowtail caterpillars turn into butterflies, put them in a jar and feed them with leaves for about two weeks.

**Cabbage Caterpillars:** These green caterpillars will eat holes in your cabbage leaves before they become pretty butterflies with white wings. Leaf holes and droppings show you where they are. Choose a cool morning when the worms are inactive and sprinkle them with rye flour or powdered lime to dehydrate their soft bodies.

**Earwigs:** They look scary but can't hurt you. They hide under leaves during the day and come out during the night to eat your weakest young plants.

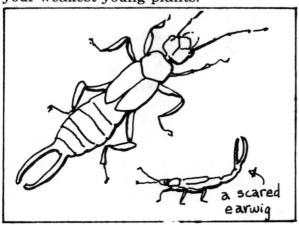

a scared earwig

**Borers:** These are unseen eaters which live inside stems, especially squash stems. They live as larvae (grubs) inside the squash vine, causing the leaves to wilt. Look for soft "soil" deposits on the stem as a clue to their presence. Slit the stem lengthwise, remove the larvae, and cover the stem with soil. To prevent eggs being laid, sprinkle black pepper around the young plants.

**Cutworms:** They eat the young plant just at the base by curling around the stem. You can trick cutworms by putting a couple of toothpicks next to the stem to change its shape. They especially like tomatoes and cabbage.

**Squash Bugs:** They suck juices from squash leaves. Use a garlic/herb spray and plant nasturtium and radish seeds in with the squash.

**Japanese Beetles:** Since they are newcomers to the United States, they have no predators yet. Sometimes they destroy a whole year's crop. When they arrive in overwhelming numbers, they can be knocked easily into a can of kerosene. The homemade repellent described on page 73 keeps them away. Starlings eat the young larvae.

**Slugs and Snails:** Tender and slimy, these feeders like moist places best. They can be killed by sprinkling them with salt or powdered lime. Attract them with bread slices or dishes of beer left out overnight.

# 43

## Life on a Carrot Flower

Most insects begin as eggs laid near a good supply of food. Sometimes these eggs produce young insects that look like their adult parents, except that they're smaller. Other times, the eggs produce young insects that look nothing like their parents. This is the case when a butterfly lays eggs that produce a caterpillar. Before a caterpillar can grow up to become a butterfly it has to go through several stages of development.

The black swallowtail butterfly (1) always lays its eggs on a plant in the carrot family. Its eggs hatch as caterpillars with green, yellow, and black stripes. The caterpillar stage of development is called the **larva** (2). The larva grows by eating leaves and sheds its skin several times. When it is ready, it changes into a **pupa** (3) or **chrysalis**. The butterfly, or **adult** form, breaks out of the chrysalis and flies away.

While a young insect is going through the egg—larva—pupa stages, it makes good food for certain other insects. The wasp (4), the praying mantis (5), and the ambush bug (6) are all hungry hunters that like to eat other insects' young. Both the ambush bug and the praying mantis are hunters all their lives: they hatch as miniature adult forms. The wasp, however, has to go through the egg—larva—pupa stages before it becomes an adult, as does the honeybee (7).

Notice the scruffy-looking beetle larva (8) behind the ladybug beetle (9). Both like to eat aphids (10), which are insects that don't lay eggs at all. Aphids give birth to live young.

This carrot flower and its insect visitors are very busy. Can you see the furry coat for pollen-gathering on the honeybee? the threatening "horns" on the frightened caterpillar? the camouflaged crab spider (11) in ambush?

# What is it?

## 44

a.

b.

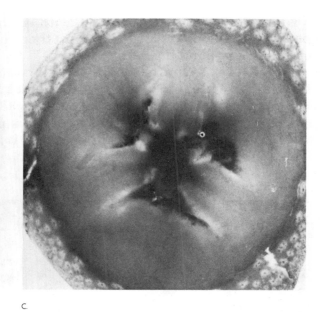

c.

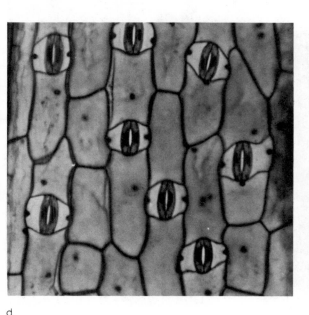

d.

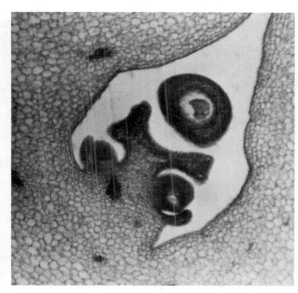

e.

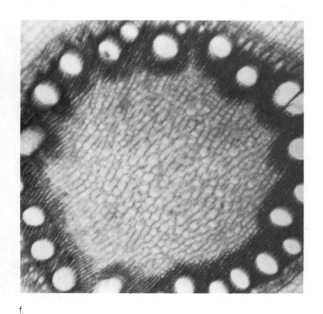

f.

Answers on copyright page, (p. 2).

## 45

# Back to Seed

We started our garden project with seeds that came in packages. All we had to do was plant them, add some water, care for them, and a new plant grew.

Now we need to find out where these seeds came from. We know seeds are made by the mature plant but until now we have not discovered exactly how a plant makes seeds. Look at the pictures of the tomato plant on this page and see if you can solve the mystery. (The seeds are inside the tomato fruit.)

Tomato Flowers

Tomato Fruit

## Flowers As Fruit and Seed Makers
## Fruits As Seed Holders

The next time you eat watermelon try to imagine it being made by a tiny flower. "Amazing!" you say. "How can a fruit that weighs up to 100 pounds come from a small flower?" Well it's true, all fruits including corn, apples, cucumbers and eggplants are made by flowers.

Open up any fruit and inside will be seeds. Fruit and seeds go together. When a flower makes a fruit, it is also making seeds. The purpose of a fruit is to hold, protect, and nourish the young seeds until they are ready to become new plants.

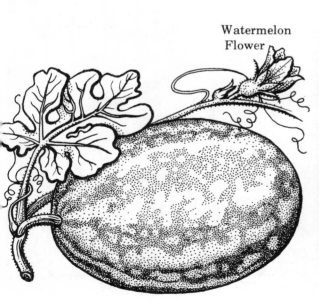

Watermelon Flower

# The Making of Fruit and Seeds

Just as the leaf contains all the equipment for making food, the flower has all the equipment for making fruit and seeds. The center picture on this page shows an apple flower cut in half so that you can see all the parts inside. Watch how the parts work together and change into fruit and seeds.

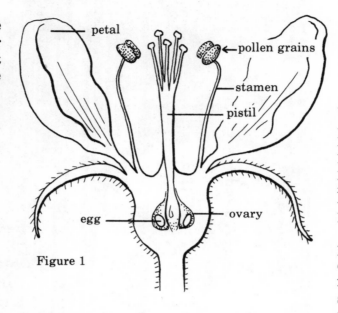

Figure 1

Apple Flower

**Figure 1** The large and usually brightly colored outer parts of a flower are the **petals**. Inside the petals are many slender threads with yellow or orange sacs at the top. These threads are called **stamens**. On each sac are a great many small grains. These grains, which look like powder or dust, are called **pollen**. The stamens form a circle around a green tube called a **pistil**. Some flowers have only one pistil, others have many. At the bottom of the pistil is a case called an **ovary**. The ovary contains **eggs**.

For a plant to make fruit and seeds, pollen must join with the eggs in an ovary of the same type plant. For instance, only apple pollen grains can join with apple eggs to make apple seeds.

When pollen joins with an egg we say the plant has been **fertilized**. For most plants to be fertilized, pollen from one plant must travel to the eggs of another plant of the same type. What would happen if pollen from a watermelon plant reached the eggs of an apple plant?

Some plants use the wind to carry their pollen to another plant, but most plants depend on our insect friends to do this important job. Let's see how an insect such as a bee helps in the joining of pollen and egg.

Apple Flower

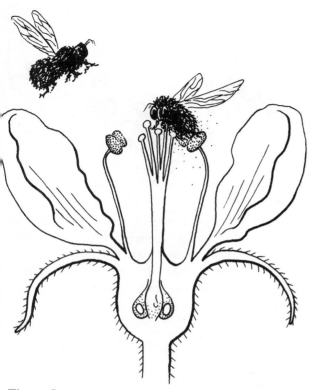

Figure 2

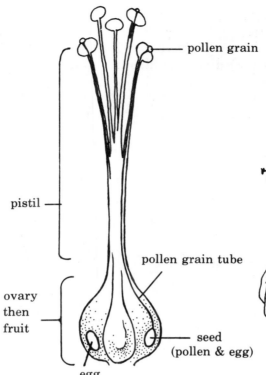

Figure 3

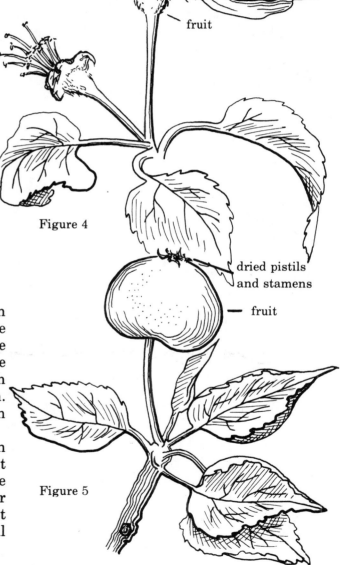

Figure 4

Figure 5

**Figure 2** Brightly colored petals, a perfumed scent, pollen, and some sweet juice called **nectar** attract the bee to the flower. The nectar and pollen will become food for the bee and its young. Do you know what nectar is made into?

The bee drinks the sweet nectar. It rubs against the stamens and collects pollen. The bee packs the pollen into tiny balls on its legs. Sometimes its entire body gets dusted with pollen as it works. The bee then flies to another flower to get more nectar and pollen. When it lands, some of the pollen from its body falls on a sticky, moist pistil.

**Figure 3** As soon as one pollen grain lands on a pistil, a miracle begins — the pollen grain starts to grow down inside the pistil and into the ovary. Once inside the ovary, it joins with and fertilizes an egg. From this joining a seed in born. Immediately the ovary begins to thicken and grow into the fruit.

**Figures 4 -7** For many days and even months, the newborn seeds and fruit grow together. In the meantime, the stamens and pistils, which are no longer of any use, dry up. The growing fruit holds and protects the young seeds until they are ready to become new plants.

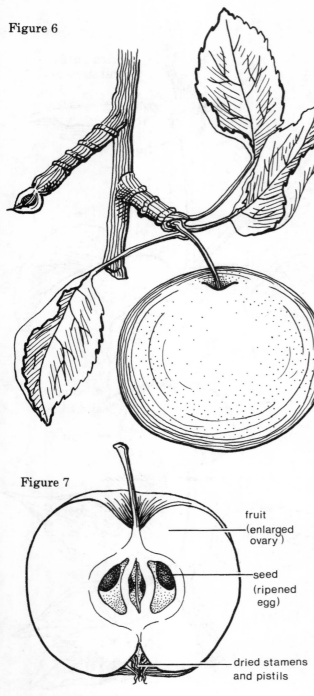

Figure 6

Figure 7
- fruit (enlarged ovary)
- seed (ripened egg)
- dried stamens and pistils

## DOING

You will need:
- magnifying glass

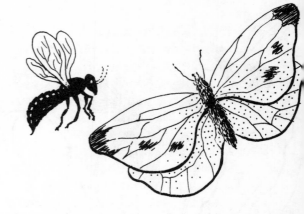

4. Watch for the insects which carry pollen for your plants. One scientist found that more than 5,000 insects visited one flower in a single day. Look closely at how insects collect nectar and pollen.

5. Read about the lives of bees and the making of honey.

1. Go into the garden and look for vegetable flowers; peas and beans are excellent. Bring a few into class, open them, and find the parts for making seeds. Use a magnifying glass. Draw a picture of what you see inside. Flowers from different vegetables will look different. How many kinds can you discover?

2. Watch as the flowers in the garden change into fruits and seeds. Open them at different stages of their growth.

3. Bring some fruits into class from home. Find the seeds in the fruit. Can you tell where the petals and pistils were? Remember, fruits are enlarged ovaries.

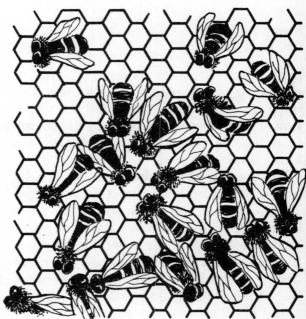

6. Every tree, weed, and vegetable has a flower. Go outside in your backyard or neighborhood. See if you can find any flowers, fruit, or seeds. Keep your eyes open and soon you'll see flowers everywhere.

7. What do these sentences mean?

"The future of a flower is a seed."

"A fruit is like a home."

"Pollen is a passenger on a trip."

8. Look at the list below. Which are flowers? Which are fruit and seeds? Which are roots, stems, or leaves?

Tomato
Lettuce
Onion
Peach
Apple
Eggplant
Carrot
Wheat
Corn
Broccoli
Pea
Bean
Cauliflower
Peanut
Banana
Rice
Squash
Acorn
Walnut

# From Outside to Inside • Harvest

## 47

Think of the word "harvest" for a moment and what it means to you. Close your eyes and feel and smell and see the time of harvesting. The harvest is the gathering of all the roots, stems, leaves, and fruits that provide food for people, and the giving of thanks to nature for having made it possible.

Make a harvest feast with your class and invite others to join. Sharing is traditional at this time.

**Till Another Year**

We have now come full cycle in our work with the garden. You have seen the seed go from germination to a plant, to a flower, to a fruit, and back again to seed.

You have seen some of the things that work well with plants and some that don't.

Finally, you have seen and experienced the importance of working with nature. Help to create a world of harmony with the knowledge you have gained.